OXFORD ENGINEERING SCIENCE SERIES

EMERITUS EDITORS

A. L. CULLEN L. C. WOODS

GENERAL EDITORS

J. M. BRADY C. E. BRENNEN W. R. EATOCK TAYLOR
M. Y. HUSSAINI T. V. JONES J. VAN BLADEL

THE OXFORD ENGINEERING SCIENCE SERIES

1 D. R. RHODES: *Synthesis of planar antenna sources*
3 R. N. FRANKLIN: *Plasma phenomena in gas discharges*
5 H.-G. UNGER: *Planar optical waveguides and fibres*
7 K. H. HUNT: *Kinematic geometry of mechanisms*
8 D. S. JONES: *Methods in electromagnetic wave propagation* (Two volumes)
10 P. HAGEDORN: *Non-linear oscillations* (Second edition)
11 R. HILL: *Mathematical theory of plasticity*
12 D. J. DAWE: *Matrix and finite element displacement analysis of structures*
13 N. W. MURRAY: *Introduction to the theory of thin-walled structures*
14 R. I. TANNER: *Engineering rheology*
15 M. F. KANNINEN and C. H. POPELAR: *Advanced fracture mechanics*
16 R. H. T. BATES and M. J. McDONNELL: *Image restoration and reconstruction*
18 K. HUSEYIN: *Multiple-parameter stability theory and its application*
19 R. N. BRACEWELL: *The Hartley transform*
20 J. WESSON: *Tokamaks*
21 P. B. WHALLEY: *Boiling, condensation, and gas–liquid flow*
22 C. SAMSON, M. LeBORGNE, and B. ESPIAU: *Robot control: the task function approach*
23 H. J. RAMM: *Fluid dynamics for the study of transonic flow*
24 R. R. A. SYMS: *Practical volume holography*
25 W. D. McCOMB: *The physics of fluid turbulence*
26 Z. P. BAZANT and L. CEDOLIN: *Stability of structures: principles of elastic, inelastic, and damage theories*
27 J. D. THORNTON: *Science and practice of liquid–liquid extraction* (Two volumes)
28 J. VAN BLADEL: *Singular elctromagnetic fields and sources*
29 M. O. TOKHI and R. R. LEITCH: *Active noise control*
30 I. V. LINDELL: *Methods for electromagnetic field analysis*
31 J. A. C. KENTFIELD: *Nonsteady, one-dimensional, internal, compressible flows*
32 W. F. HOSFORD: *Mechanics of crystals and polycrystals*
33 G. S. H. LOCK: *The tubular thermosyphon: variation on a theme*
34 A. LIÑÁN and F. A. WILLIAMS: *Fundamental aspects of combustion*
35 N. FACHÉ, F. OLYSLAGER, and D. DE ZUTTER: *Electromagnetic and circuit modelling of multiconductor transmission lines*

Electromagnetic and Circuit Modelling of Multiconductor Transmission Lines

NIELS FACHÉ
FRANK OLYSLAGER
DANIËL DE ZUTTER

Laboratorium voor Elektromagnetisme en Acustica
Universiteit Gent

CLARENDON PRESS · OXFORD

This book has been printed digitally and produced in a standard specification in order to ensure its continuing availability

OXFORD
UNIVERSITY PRESS

Great Clarendon Street, Oxford OX2 6DP

Oxford University Press is a department of the University of Oxford.
It furthers the University's objective of excellence in research, scholarship,
and education by publishing worldwide in

Oxford New York

Auckland   Cape Town Dar es Salaam Hong Kong   Karachi
Kuala Lumpur Madrid Melbourne Mexico City   Nairobi
New Delhi Shanghai Taipei Toronto
With offices in
Argentina Austria Brazil Chile Czech Republic France Greece
Guatemala Hungary Italy Japan South Korea Poland Portugal
Singapore Switzerland Thailand Turkey Ukraine Vietnam

Oxford is a registered trade mark of Oxford University Press
in the UK and in certain other countries

Published in the United States
by Oxford University Press Inc., New York

© Niels Faché, Frank Olyslager, and Daniël De Zutter, 1993

The moral rights of the author have been asserted

Database right Oxford University Press (maker)

Reprinted 2011

All rights reserved. No part of this publication may be reproduced,
stored in a retrieval system, or transmitted, in any form or by any means,
without the prior permission in writing of Oxford University Press,
or as expressly permitted by law, or under terms agreed with the appropriate
reprographics rights organization. Enquiries concerning reproduction
outside the scope of the above should be sent to the Rights Department,
Oxford University Press, at the address above

You must not circulate this book in any other binding or cover
And you must impose this same condition on any acquirer

ISBN 978-0-19-856250-4

To my parents, Irene and André
NIELS FACHÉ

In memory of my grandfather, J. B. Leo
FRANK OLYSLAGER

To my wife, Inge.
DANIËL DE ZUTTER

PREFACE

This book contains two parts. The first part deals with circuit modelling of multiconductor waveguides while the second part discusses the full-wave electromagnetic modelling of a particular subclass of these waveguides. Although there is a strong relationship between both topics, they are presented in such a way that the two parts of the book can be read almost independently. To that end a separate introductory chapter has been written for each of the two topics: Chapter 1 introduces the main principles of multiconductor circuit modelling while Chapter 5 discusses the waveguide structures analysed in this book and the full-wave techniques used to perform that analysis. The first section of Chapter 1 provides a more general introduction to the subject matter of the book.

The increasing use of microwave and millimeter-wave circuits and the introduction of high-speed digital communication systems have substantially increased research on the modelling of interconnections and of circuit elements with distributed parameters. Powerful CAD packages have emerged to design microwave systems, and digital routing tools are now extended in order to take the high-speed behaviour of the interconnections into account.

This book contributes to the study of multiconductor transmission lines. They form the basic building block of microwave and millimeter-wave integrated circuits and are omnipresent in digital systems. For circuit and network simulation, good models of these multiconductors lines are indispensable.

Chapters 1–4 give a detailed account of the way in which such models can be obtained starting from a full-wave modal description of multiconductor waveguides. Special emphasis is placed on the principles and restrictions involved in the transition from the electromagnetic field description to the circuit description. These chapters cover most of the knowledge in the field and offer the reader the latest advances with respect to the circuit modelling of coupled lossy lines.

Chapters 5–9 give a detailed account of the full-wave eigenmode analysis of several types of multiconductor waveguides embedded in a planar stratified medium. In each case the solution of the problem is based on a suitable integral equation technique. For infinitely thin coupled microstrips and striplines this integral equation is solved in the space domain. For discrete wire structures the solution is partly effected in the space domain, partly in the spectral domain. Finally, the case of coupled lossy polygonal conductors is handled in the spectral domain. In all cases special attention is devoted to the impedance calculations. Together with the propagation constants of the modes, these impedances are necessary to construct a circuit model. A separate chapter is also devoted to spectral domain calculations in multilayered media.

The authors are indebted to Professor Paul Lagasse, Director of the Laboratory of Electromagnetism and Acoustics of the University of Ghent, for providing the optimum intellectual and material conditions for the research presented in this book. They are grateful to Professor Jean Van Bladel for the suggestion and encouragement to undertake the writing of this book.

We specifically want to thank Tom Dhaene who was responsible for the basic work presented in Chapter 4. The contributions of Jan Van Hese to Chapter 7 and of Krist Blomme to Chapter 9 are gratefully acknowledged. We also very much appreciated the team spirit within the high-speed/high-frequency group of the Laboratory of Electromagnetism and Acoustics.

Finally, the authors acknowledge the support of the National Fund for Scientic Research of Belgium (NFWO). Frank Olyslager is a Research Assistant and Professor Daniël De Zutter is a Research Director of the NFWO.

Ghent N.F.
April 1992 F.O.
  D.D.Z.

CONTENTS

LIST OF SYMBOLS — XV

1 GENERAL INTRODUCTION TO THE CIRCUIT MODELLING OF MULTICONDUCTOR LINES — 1
  1.1 Overview of the electromagnetic and circuit modelling problem — 1
  1.2 Circuit modelling of multiconductor lines — 3
    1.2.1 Introduction — 3
    1.2.2 Classical circuit description — 4
    1.2.3 Electromagnetic versus circuit description of a multiconductor line — 5
    1.2.4 From a field description to a circuit description — 6
    1.2.5 Power equivalence and circuit principle — 7
    1.2.6 TEM waveguide modes and consistency with the classical circuit description — 8
    1.2.7 Power–current and power–voltage model — 9
    1.2.8 Reciprocity of the circuit model — 11
    1.2.9 Characteristic impedance matrix, impedance and admittance matrix, line-mode impedances — 12
  References — 13

2 CIRCUIT DESCRIPTION OF SINGLE WAVEGUIDE STRUCTURES — 16
  2.1 Introduction — 16
  2.2 Modal field representation — 17
    2.2.1 TEM mode — 17
    2.2.2 Hybrid mode — 22
  2.3 Application of the power equivalence principle — 26
    2.3.1 TEM mode — 26
    2.3.2 Hybrid mode — 28
  2.4 Application of the circuit principle — 32
    2.4.1 Introduction — 32
    2.4.2 Circuit models for load and driver — 33
    2.4.3 Calculation of the amplitude of the characteristic impedance — 36
  References — 51

3 CIRCUIT DESCRIPTION OF TWO COUPLED LOSSLESS WAVEGUIDE STRUCTURES — 53
  3.1 Introduction — 53
  3.2 Two coupled lossless hybrid waveguide structures — 53
  3.3 Modal field representation — 54
  3.4 Coupled equivalent transmission lines — 58
  3.5 Circuit model — 63
    3.5.1 Circuit currents — 63

3.5.2	Circuit line-mode impedances	65
3.5.3	Circuit voltages	67
3.5.4	Circuit equations	67
3.5.5	Power and field distribution in the circuit	68
References		74

4 CIRCUIT DESCRIPTION OF GENERAL LOSSY MULTI-CONDUCTOR WAVEGUIDE STRUCTURES — 75

- 4.1 Introduction — 75
- 4.2 General coupled waveguide structure — 76
- 4.3 Modal field representation — 76
- 4.4 Transformation from a modal to a circuit description — 78
  - 4.4.1 General formulation — 78
  - 4.4.2 Circuit current — 78
  - 4.4.3 Circuit voltage — 79
  - 4.4.4 Power conservation — 80
  - 4.4.5 Arbitrariness of M_I and M_V — 80
  - 4.4.6 Power–voltage circuit model — 81
- 4.5 Coupled transmission line model — 82
  - 4.5.1 Generalized telegrapher's equations — 82
  - 4.5.2 Characteristic impedance versus line-mode impedance — 85
  - 4.5.3 Reciprocity of the full-wave circuit model — 87
- 4.6 Case study — 89
- References — 94

5 GENERAL INTRODUCTION TO THE FULL-WAVE ANALYSIS OF MULTICONDUCTOR LINES IN A PLANAR STRATIFIED MEDIUM — 95

- 5.1 Full-wave analysis of two- and three-dimensional waveguide structures embedded in a planar stratified medium — 95
- 5.2 Infinitely thin strip conductor lines — 97
- 5.3 Spectral domain description of a planar stratified medium — 99
- 5.4 Discrete wire lines — 100
- 5.5 Multiconductor lines formed by polygonal-shaped conductors with finite conductivity — 103
- 5.6 Additional features of the integral equation approach — 107
  - 5.6.1 Modelling of the singular behaviour of the current near conductor edges — 107
  - 5.6.2 Extended spectral domain approach — 108
- References — 109

6 SPECTRAL FIELD CALCULATIONS IN A MULTI-LAYERED MEDIUM — 112

- 6.1 Introduction — 112
- 6.2 The electromagnetic field in the spectral domain — 114
  - 6.2.1 The spatial electromagnetic field in a sourceless layer — 115
  - 6.2.2 The spectral transformation of the electromagnetic field — 116

	6.2.3	General representation of the spectral fields: TE and TM modes	117
	6.2.4	The transmission line cascade model for the TE and TM modes	120
	6.2.5	Transformation formulas in the TE and TM cascade	121
	6.2.6	Solution of the TE and TM cascade	124
	6.2.7	Numerical considerations	125
6.3		The electromagnetic field in the space domain	130
6.4		Power propagated in a sourceless layer	134
References			136

7 INFINITELY THIN STRIP TRANSMISSION LINE STRUCTURES — 137

- 7.1 Introduction — 137
- 7.2 Geometry of the strip structures — 138
- 7.3 Construction of the integral equation — 139
- 7.4 Method of moments solution technique — 141
  - 7.4.1 Current basis functions — 141
  - 7.4.2 Testing of the integral equation — 142
  - 7.4.3 Solution of the discrete eigenvalue problem — 143
- 7.5 The spectral Green's dyadic — 145
  - 7.5.1 The excitation of the TE and TM cascade — 145
  - 7.5.2 Solution of the TE and TM cascade — 146
  - 7.5.3 Construction of the Green's dyadic — 147
- 7.6 The spatial Green's dyadic — 149
  - 7.6.1 The inverse Fourier transformation — 149
  - 7.6.2 The evaluation of the TE and TM integrals — 150
- 7.7 Calculation of the contribution of each basis function to the integral equation — 155
  - 7.7.1 Non-self patch contributions — 156
  - 7.7.2 Selfpatch contributions — 157
- 7.8 Case studies — 159
  - 7.8.1 A single microstrip — 159
  - 7.8.2 A suspended substrate line — 161
  - 7.8.3 Two coupled symmetric microstrips — 162
  - 7.8.4 Two coupled asymmetric microstrips — 164
  - 7.8.5 Three coupled lines — 166
- References — 167

8 LOSSLESS WIRE TRANSMISSION LINE STRUCTURES — 169

- 8.1 Introduction — 169
- 8.2 Geometry — 170
- 8.3 Construction of the integral equation — 171
  - 8.3.1 General representation of the modal fields — 171
  - 8.3.2 The integral equation — 172
- 8.4 Solution of the integral equation — 173
  - 8.4.1 The incoming fields — 174

	8.4.2	The scattered fields	177
	8.4.3	Construction of the discrete eigenvalue problem	178
8.5	The incoming fields		179
	8.5.1	The spatial incoming fields	179
	8.5.2	The spectral incoming fields	180
	8.5.3	Angular decomposition of the incoming fields on the wire surfaces	185
8.6	The scattered fields		187
	8.6.1	The spectral scattered fields	187
	8.6.2	The spatial scattered fields and the angular decomposition on the wire surfaces	188
8.7	The inverse Fourier transformation		190
8.8	The longitudinal currents		192
8.9	The propagated power		192
	8.9.1.	Introduction	192
	8.9.2	Power propagated in a layer with wires	193
8.10	Case studies		195
	8.10.1	A single wire in a two-layered medium with ground plane	195
	8.10.2	Two wires in a double-layered medium with ground plane	198
	8.10.3	Two wires in a multilayered medium	200
References			201

9 ARBITRARILY-SHAPED POLYGONAL TRANSMISSION LINE STRUCTURES WITH FINITE CONDUCTIVITY 203

9.1	Introduction		203
9.2	Geometry		204
9.3	Construction of the integral equation		206
	9.3.1	Introduction	206
	9.3.2	Internal regions	206
	9.3.3	External regions	207
	9.3.4	Boundary conditions and unknowns	208
	9.3.5	Integral equation	209
	9.3.6	Perfectly conducting conductors	210
9.4	Solution technique for the integral equation		211
	9.4.1	Basis functions and test functions	211
	9.4.2	Discrete eigenvalue problem	213
	9.4.3	The system matrix	214
	9.4.4	Incoming coefficients	217
	9.4.5	Scattered coefficients	218
9.5	Excitation and observation integrations		219
	9.5.1	Polygonal conductors	219
	9.5.2	Excitation integration	219
	9.5.3	Observation integration	222
9.6	Inverse Fourier transformation		224
9.7	Impedance calculation		226
	9.7.1	Introduction	226

	9.7.2 Longitudinal current	226
	9.7.3 Propagated power	227
9.8	Case studies	230
	9.8.1 Microstrip configuration	230
	9.8.2 Wire transmission lines	233
	9.8.3 A multiconductor multilayer stripline	237
References		239

APPENDIX A
Integral representation of the longitudinal modal fields   241

AUTHOR INDEX   247

SUBJECT INDEX   249

LIST OF SYMBOLS

The imaginary symbol is j, and the harmonic time factor is $e^{j\omega t}$.
The chapter mentioned between the brackets refers to the chapter in which the quantity is defined or extensively used.

a	radius of a wire (Chapter 8)
A	wave coefficient in spectral TE or TM transmission line cascade (Chapters 6–9)
B	wave coefficient in spectral TE or TM transmission line cascade (Chapters 6–9)
B	magnetic induction
c	used as a subscript, denotes a circuit transmission line quantity
C	capacitance per unit of length (Chapter 2)
C	total number of polygonal conductors (Chapter 9)
C	column vector formed by the spectral electric and magnetic fields (Chapter 6)
C	capacitance matrix (Chapter 4)
d	substrate thickness (Chapters 7 and 8)
D	total numbers of wires (Chapter 8)
D	used as a subscript, denotes a driver quantity (Chapter 2)
D	electrical induction
e	used as a subscript, denotes an equivalent field transmission line quantity (Chapters 2 and 3)
e	used as a subscript, denotes excitation strips, wires, polygonal conductors, or layers (Chapters 6–9)
e	electric field with explicit x and/or time dependence (Chapters 6 and 7, and Appendix A)
Ei(z)	exponential integral of argument z (Chapter 7)
E	electric field
$E_3(z)$	exponential integral of argument z and order 3 (Chapter 9)
f	frequency
f	used as a subscript, denotes a field transmission line quantity (Chapters 2–4)
G	part of the electric Green's dyadic $\mathbf{G}_e$ relevant to the tangential electric field components (Chapter 7)
$G(\mathbf{r}\|\mathbf{r}')$	two-dimensional Green's function of layer j (Chapters 8 and 9, and Appendix A)
G_{ab}	one of the four components of the Green's dyadic **G** ($a = x$ or y, $b = x$ or y) (Chapter 7)

G	conductance matrix (Chapter 4)
$\mathbf{G}_e$	electric Green's dyadic (Chapter 7)
G	electric spectral domain Green's function (Chapter 7)
h	substrate thickness (Chapter 9)
$\mathbf{h}$	magnetic field with explicit x and/or time dependence (Chapters 6 and 7, and Appendix A)
H	distance between ground plane and centre of wire (Chapters 8 and 9)
$\mathbf{H}$	magnetic field
$H_n^{(2)}(z)$	second-kind Hankel function of argument z and order n (Chapters 8 and 9)
(i)	used as a superscript, denotes a quantity belonging to mode i (Chapter 3)
i_j	used as a subscript, refers to the jth wire or the jth polygonal conductor in layer i (Chapters 8 and 9)
in	used as superscript on field quantities, denotes incoming fields (Chapters 8 and 9)
I	denotes a current, in most cases used with various subscripts and/or superscripts
$\mathbf{J}_j$	surface current density on thin strip conductor j (Chapter 7)
$J_n(z)$	first-kind Bessel function of argument z and order n (Chapter 8)
$J_x(y)$	x-component of the surface current density (Chapter 7)
$J_y(y)$	y-component of the surface current density (Chapter 7)
$\mathbf{k}$	$(=\beta\mathbf{1}_x + k_y\mathbf{1}_y)$ wave vector (Chapter 6)
k^2	$(=\mathbf{k}\cdot\mathbf{k})$ square of the amplitude of the wave vector $\mathbf{k}$ (Chapter 6)
k_b	branch point in the complex k_y plane (Chapter 6)
k_c	limiting value of the asymptotic range of k_y values (Chapter 6)
k_y	$(=k_y' + jk_y'')$ spectral variable (Chapter 6)
k_0	$(=\omega\sqrt{\varepsilon_0\mu_0})$ wave number of vacuum
k_p	pole in the complex k_y plane (Chapter 6)
l	used as a subscript, denotes a longitudinal field component (Chapters 2–4)
lm	used as a subscript, denotes a line-mode quantity (Chapters 3 and 4)
L	inductance per unit of length (Chapter 2)
L	total number of layers in a multilayered medium (Chapter 6)
L	used as a subscript, denotes a load quantity (Chapter 2)
L	inductance matrix (Chapter 4)
m_0	shorthand for the expression defined in equation (9.35)
$\mathbf{M}_i^u$	up-transformation matrix in the spectral domain (Chapter 6)
$\mathbf{M}_i^d$	down-transformation matrix in the spectral domain (Chapter 6)
M_I	transformation matrix between modal currents and circuit currents (Chapter 4)

LIST OF SYMBOLS

M_V	transformation matrix between modal voltages and circuit voltages (Chapter 4)
n	used as a subscript to denote a normal field component (Chapter 9)
n	unit normal vector to a boundary curve (Chapter 9)
n_0	shorthand for the expression defined in equation (9.37)
o	used as a subscript, denotes observation strips, wires, polygonal conductors, or layers (Chapters 6–9)
$p(\tau)$	pulse basis function on the boundary of a polygonal conductor (Chapter 9)
p	power propagated by the modal electric field of one mode and the modal magnetic field of another or of the same mode, in many cases used with various subscripts and/or superscripts (Chapters 2–4)
P	a propagated power, in most cases used with various subscripts
Q	charge per unit of length (Chapter 2)
r	distance to the origin in a polar coordinate system
R	resistance matrix (Chapter 4)
R_L	unknown wave coefficient in spectral cascade (Chapter 6)
R_1	unknown wave coefficient in spectral cascade (Chapter 6)
R_0	$(=\sqrt{\mu_0/\varepsilon_0})$ characteristic impedance of vacuum
s	distance between conductors (Chapters 7–9)
sc	used as superscript on field quantities, denotes scattered fields (Chapters 8 and 9)
sign(z)	signum function of argument z (Chapter 7)
$S(\tau)$	Fresnel sine integral of argument τ (Chapter 7)
t	time coordinate
t	used as a subscript, denotes a transversal field component (Chapters 2–4)
t	used as a subscript, denotes a tangential field component (Chapter 9)
t	shorthand for the expression defined in equation (9.35)
t	conductor thickness (Chapter 9)
t	tangent vector (Chapter 9)
tg δ	loss tangent (Chapter 9)
$t(y')$	generic notation for a basis function on a thin strip conductor (Chapter 7)
$t(\tau)$	triangular basis function on the boundary of a polygonal conductor (Chapter 9)
T	used as a superscript, denotes the transposition operator (Chapter 4)
TE	used as a subscript, denotes TE mode quantities (Chapters 6–9)
TM	used as a subscript, denotes TM mode quantities (Chapters 6–9)
u	shorthand for the expression defined in equation (9.37)
v	shorthand for the expression defined in equation (9.70)
V	a voltage, in most cases used with various subscripts and/or superscripts

LIST OF SYMBOLS

w	conductor width (Chapters 7–9)
w	shorthand for the expression defined in equation (9.42)
x	coordinate in the propagation direction
y	coordinate in the cross-section of the waveguide
y	admittance in telegrapher's equations, used with various subscripts (Chapters 3 and 4)
z	coordinate in the cross-section of the waveguide
z	impedance in telegrapher's equations, used with various subscripts (Chapters 3 and 4)
Z	an impedance, in most cases used with various subscripts and/or superscripts
α	angle of basis function segment with y-axis (Chapter 9)
β	$(=\beta' - j\beta'')$ complex propagation constant of a mode, in several cases used with various subscripts and/or superscripts
γ	$(=-\gamma' - j\gamma'' = \sqrt{\omega^2 \varepsilon \mu - \beta^2})$ (Chapters 6–9)
Γ	spectral propagation constant (Chapter 6)
δ	arbitrary multiplication factor (Chapters 2–4)
δ	difference in y-coordinate between excitation point and matching point (Chapter 7)
δ	length of a basis function interval (Chapter 9)
Δ	determinant of a matrix (Chapter 3)
Δ	length of a basis function interval on a thin strip conductor (Chapter 7)
Δy	distance between centre of observation wire and excitation wire (Chapter 8)
Δz	difference in z-coordinate between excitation and observation interval (Chapter 7) or between centre of observation wire and excitation wire (Chapter 8)
$\varepsilon, \varepsilon_r$	(complex) permittivity and relative permittivity
$\varepsilon_{r,eff}$	effective permittivity (Chapter 7)
ε_0	permittivity of vacuum
μ, μ_r	(complex) permeability and relative permeability
μ_0	permeability of vacuum
ν	distance coordinate along a conductor boundary (Chapter 9)
ξ	reciprocity check variable (Chapter 3)
σ	conductivity
τ	distance coordinate along a conductor boundary (Chapters 7 and 9)
ϕ	angle in a polar coordinate system
ϕ	electric field potential (Chapter 4)
Φ	flux of magnetic induction per unit of length (Chapter 2)
χ	constant of separation of variables (Chapter 2)
ψ	magnetic field potential (Chapter 4)

LIST OF SYMBOLS

ω	angular frequency or pulsation
∇_t	transversal nabla operator ($= \partial/\partial y \, \mathbf{1}_y + \partial/\partial x \, \mathbf{1}_z$)
$\mathbf{1}_a$	unit vector in the a-direction
$\partial/\partial t$	tangential derivative (Chapter 9)
$\partial/\partial n$	normal derivative (Chapter 9)
$+$	used as a superscript, refers to a wave propagating in the positive x-direction (Chapters 2–4) or in the positive z-direction (Chapters 8 and 9)
$-$	used as a superscript, refers to a wave propagating in the negative x-direction (Chapters 2–4) or in the negative z-direction (Chapters 8 and 9)
$'$	used as a superscript, denotes the component of a vector in the spectral domain in the direction of $\mathbf{k}$ (Chapter 6)
$''$	used as a superscript, denotes the component of a vector in the spectral domain in the direction of $\mathbf{1}_z \times \mathbf{k}$ (Chapter 6)
$*$	used as a superscript, denotes the complex conjugate operator

1
GENERAL INTRODUCTION TO THE CIRCUIT MODELLING OF MULTICONDUCTOR LINES

This book is organized in two parts. The first part (Chapters 1–4) deals with the circuit modelling of multiconductor waveguides. The second part (Chapters 5–9) deals with the full-wave electromagnetic modelling of some particular subclasses of these waveguides. The present chapter starts with a short general introduction to both problems and is followed by a much more extensive introduction to the circuit modelling problem. An extensive introduction to the electromagnetic field problems treated in this book is given in Chapter 5.

1.1 Overview of the electromagnetic and circuit modelling problem

Multiconductor transmission line structures form a basic building block of microwave integrated circuits. The prediction of their behaviour at microwave and millimeter-wave frequencies is essential to the correct functioning of these devices. For Monolithic Microwave Integrated Circuits (MMICs) a first pass design is set as a goal. Several commercial CAD packages are available for the analysis of the behaviour of complex circuits [1]. They include the use of such elements as spiral transformers, inductors, interdigital finger capacitors and filters, to name but a few. These CAD packages are based on the availability of the circuit parameters (either S-parameters, Y-parameters, or some other equivalent representation) of a large number of basic building blocks such as interconnecting lines, discontinuities, and active circuits [1]. Multiconductor transmission line sections are basic modules for the layout-oriented simulation of various coupling situations and of various lumped elements [2]. It is important to realize however that the circuit approach which forms the basis of most large CAD packages relies on the TEM description of its basic building blocks. This implies that the description of multiconductor transmission lines, such as (coupled) microstrips, slotlines, or coplanar waveguides, which are essentially non-TEM structures, must be accommodated in one way or another by such a TEM description. The first part of this book is devoted to that particular problem and a more detailed introduction to this problem and to Chapters 2, 3, and 4 can be found below in Section 2.

1 GENERAL INTRODUCTION

The presence of multiconductor transmission lines is certainly not restricted to Microwave Integrated Circuits (MICs) and MMICs. Due to the increasing bit-rates in emerging digital broadband systems and in computers, the behaviour of multiconductor buses on boards and backplanes becomes increasingly important. Here again, the prediction of digital signal behaviour includes not only transmission, reflection, and crosstalk in buses but also the behaviour of the active circuits, packages, connectors, pads, and via-holes and other discontinuities on the signal propagation path. This is possible only if powerful network simulation programs are used (such as the many variants of the well-known SPICE program) and here again a circuit description of multiconductor buses is necessary.

Multiconductor lines or buses are structures with a constant cross-section which propagate a number of modes. Each of these solutions to the sourceless Maxwell's equations is characterized by its modal propagation constant and by the modal field distribution in the cross-section. In the direction perpendicular to the cross-section, that is, the propagation direction, the (complex) field amplitudes essentially behave as voltage and current on a transmission line. It is quite self-evident that the equivalent circuit model for these multiconductor lines turns out to be a set of coupled transmission lines. However, the relation between the typical circuit quantities, such as voltages, currents, (coupling) impedances and admittances, and signal velocities, on the one hand, and, on the other hand, the original field quantities, that is the modal fields and associated modal propagation constants, is not straightforward. A prerequisite for the determination of the parameters of any circuit model is that the associated field problem has to be solved. The second part of this book, Chapters 5–9, is devoted to the solution of this field problem for three specific types of multiconductor lines: infinitely thin perfectly conducting strips, perfectly conducting wire transmission lines, and arbitrarily shaped polygonal conductors with finite conductivity. These multiconductor lines are embedded in a very general planar stratified medium. In each case a full-wave analysis will be presented, the results of which can be used to determine the frequency-dependent circuit parameters of an equivalent circuit model. The numerical data for the examples presented in Chapters 2, 3, and 4 to illustrate our approach to the circuit modelling of multiconductor lines, are obtained with software packages based on the theory presented in the second part of the book.

At this point two additional remarks must be made. The circuit modelling part of the book deals with very general multiconductor waveguide structures and is not restricted to those types for which the full-wave analysis is presented in the second part of the book. Furthermore, it is important to note that the full-wave and circuit analyses are restricted to the frequency domain. It is clear however that the frequency domain circuit model parameters can be used as the starting point for the time-domain analysis of networks including multiconductor lines. This is a vast research topic with important technical applications. One essentially uses (fast) Fourier transform techniques to transform the frequency domain data to the time domain and to calculate the response of the network by a convolution approach. Some of the work presented in this book has been used for that purpose [3]–[9].

For a general introduction to the full-wave analysis of the multiconductor lines treated in the second part of this book, the reader is referred to Chapter 5. The remaining part of this introductory chapter is devoted to the subject matter of the first part of the book, that is, the circuit modelling of multiconductor lines.

1.2 Circuit modelling of multiconductor lines

1.2.1 *Introduction*

As stated above, the main problem that will be addressed in Chapters 2–4 is the representation of a set of coupled multiconductor waveguides, essentially carrying a number of hybrid modes, by an 'equivalent' circuit representation which can serve as a building block in CAD software. The term 'hydrid' means that the electric and magnetic fields of the mode have both components in the cross-section of the waveguide and along the propagation direction perpendicular to that cross-section. The meaning of the word equivalent will be discussed in the sequel. The literature on the subject is quite abundant and we will not try to reconstruct an extensive picture of the different attempts that have been made to solve the problem. While presenting the major lines of thought put forward in this book, we will mention the most relevant contributions to this research topic to be found in the literature, but we do not claim to be complete. A number of additional references can be found in subsequent chapters.

Chapters 2–4 present a consistent theory of the circuit modelling of multiconductor lines. In Chapter 2 the circuit model for a single waveguide structure is discussed, followed by the derivation of the circuit model for two coupled lossless waveguides in Chapter 3. Finally, Chapter 4 deals with the extension of the theory to lossy multiconductor lines. Chapter 2 is organized in such a way that readers who are not familiar with the subject matter can acquaint themselves with the different aspects of the problem. To this end we also discuss the familiar case of a simple TEM waveguide (such as a coaxial cable) in order to clarify some of the concepts and difficulties that arise while discussing general hybrid waveguide structures.

1 GENERAL INTRODUCTION

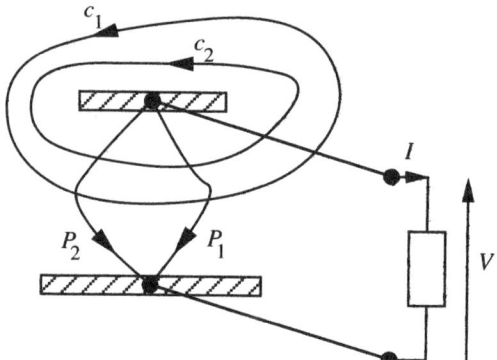

Fig. 1.1. Classical circuit definition of voltage and current.

1.2.2 *Classical circuit description*

In classical low-frequency circuit theory the voltage V and the current I are related to the electrical field **E** and magnetic field **H** at the port of a circuit element in the following way:

$$V = \int_P \mathbf{E} \cdot d\mathbf{l}$$
$$I = \oint_c \mathbf{H} \cdot d\mathbf{l} \tag{1.1}$$

where the first integral is taken from one conductor to the other and the second integral is taken around one of the conductors. The path P (for example P_1 or P_2 in Fig. 1.1) between the two conductors is arbitrary, as is the contour c (such as c_1 or c_2) around one of the conductors. This is the case for elements, junctions, and ports which are small with respect to the wavelength. However if we start from the integral form of Maxwell's curl equations (assuming an $e^{j\omega t}$ time dependence):

$$\oint_{P_{tot}} \mathbf{E} \cdot d\mathbf{l} = -j\omega \int_S \int \mathbf{B} \cdot d\mathbf{S}$$
$$\oint_c \mathbf{H} \cdot d\mathbf{l} = I + j\omega \int_A \int \mathbf{D} \cdot d\mathbf{S} \tag{1.2}$$

where P_{tot} is the total closed contour formed by joining P_1 and P_2, we can see immediately that the integrals over P_1 and P_2 in (1.1) will be equal only if the frequency ω multiplied by the flux of the magnetic induction **B** through the surface S generated by P_{tot} is negligible. An analogous reasoning applies to the definition of I in (1.1). Here the value obtained for c_1 will not differ from the one obtained for c_2 if the product of the frequency and the flux of the electric induction **D** through the surface A, generated by either c_1 or c_2, is negligible. For low frequencies and over small distances with respect to the wavelength the effect of the fluxes of **B** and **D** will indeed be negligible and hence the voltage V and the current I are defined in a unique way. In that case the average complex power P_c entering the element of Fig. 1.1 at its terminals and the impedance Z_c seen at these terminals are given by

$$P_c = \tfrac{1}{2}VI^*$$
$$Z_c = \frac{V}{I}. \tag{1.3}$$

The fact that the flux contributions in (1.2) can always be neglected at sufficiently low frequencies provides the basis for classical circuit theory.

1.2.3 *Electromagnetic versus circuit description of a multiconductor line*

From the circuit element of Fig. 1.1, we now turn to single conductor, coupled conductor, and multiconductor transmission lines. To clarify the concepts we take the example of the microstrip transmission line (see Fig. 1.2). It is our purpose to represent this structure by a circuit element. This element is the circuit transmission line of Fig. 1.3. In circuit theory, this transmission line is completely characterized by its propagation constant β, its length L, and its characteristic impedance Z_c. Voltage and current are defined at each point on the transmission line and in particular at its ports. By simply requiring Kirchhoff's voltage and current laws to be satisfied at each port, this particular circuit element can easily be connected to other circuit elements. The real microstrip transmission line structure is completely described in terms of the electric field **E** and the magnetic field **H**. However, for realistic dimensions of the microstrip and at those frequencies at which this type of transmission line is used, the flux contributions from the right-hand side of (1.2) are not negligible and equations (1.1) can no longer

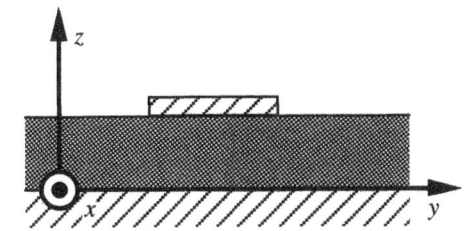

Fig. 1.2. Cross-section of a microstrip transmission line.

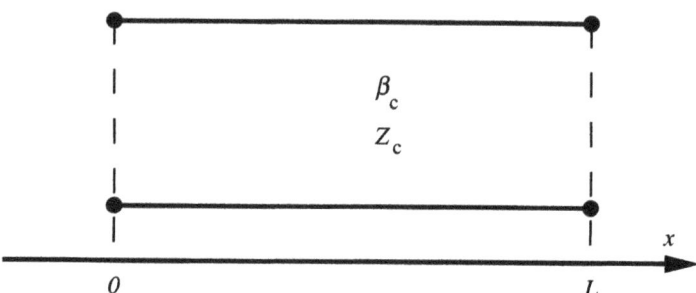

Fig. 1.3. Equivalent circuit transmission line for the structure of Fig. 1.2.

be used to define a voltage and a current unequivocally. Consequently, the discussion arises of how to connect the field quantities describing the microstrop line with the circuit quantities describing the circuit transmission line. It is quite natural to identify the propagation constant β of the circuit transmission line with the propagation constant of the fields along the microstrip. This means that sinusoidal waves or digital signals propagate with the same speed and with the same dispersion characteristics in the equivalent circuit model as in the real waveguide. The ambiguity in the definition of V and I however gave rise to a debate on the definition of the characteristic impedance Z_c. We refer the interested reader to a selection [10]–[26] of the many papers which contributed to that debate for an overview of the different ways in which the problem was tackled and to the wealth of additional references to be found in these papers.

1.2.4 From a field description to a circuit description

At the present moment it is generally accepted that any circuit model should be based on a rigorous full-wave solution of the field problem [12], [14], [18], [19]. In line with this, our description of multiconductor waveguides in subsequent chapters will start from a modal description of the electromagnetic fields. This description will be restricted to the fundamental modes propagated by the structures under study. The fundamental mode propagated by a microstrip line can be described in terms of a modal electric and a modal magnetic field together with a so-called field voltage, field current, and associated field impedance. The modal fields depend only upon the coordinates in the plane perpendicular to the propagation direction of the mode (the (y, z) plane in Fig. 1.2), while the field voltage V_f and the field current I_f depend only upon the coordinate along the propagation direction (that is, x in Fig. 1.3). V_f and I_f are related to each other by transmission-line-type equations involving Z_f and the propagation constant β of the mode. In the terminology put forward in this book this transmission line is called

1.2 CIRCUIT MODELLING OF MULTICONDUCTOR LINES

the *field transmission line*. The notion of field voltage, field current, and field impedance are common to the description of any type of waveguide. As discussed in detail in Chapter 2, a certain amount of arbitrariness in the modal field description allows one to select any complex value of Z_f. The different principles that have to be invoked and the approximations that have to be made in order to 'select' one value of Z_f as the value to be used in the equivalent circuit transmission line is what really constitutes the subject matter of the next three chapters.

1.2.5 *Power equivalence and circuit principle*

Starting from the field description, we will first invoke the *power equivalence* principle. This principle states that the average power P_f propagated by the actual waveguide must be the same as the average power P_c propagated by its equivalent circuit representation, that is $P_c = (VI^*/2)$. This assumption is of course perfectly natural as one does not want to violate the power conservation principle. Note that both the real power and the reactive power are conserved. Invoking this principle will result in fixing the phase of the field impedance Z_f. All field transmission lines satisfying the power equivalence principle and hence having the same phase value for Z_f are indicated in our terminology as *equivalent field transmission lines*.

The second and final step in selecting one single value for Z_f, to become the characteristic impedance Z_c of the circuit transmission line representation of the waveguide, is more delicate. It consists of selecting a suitable definition for either the circuit voltage or the circuit current. In view of the inherent ambiguity of such definitions in general, this amounts to selecting one particular path P or one particular contour c in equation (1.1). In the sequel this selection of a definition for either V or I will be referred to as the *circuit principle*. In order to respect the power equivalence principle introduced above, it will be shown in Chapter 2 that it is not possible, and indeed erroneous, to define both a voltage and a current at the same time. Definition of either voltage or current automatically fixes the amplitude of the field impedance Z_f. By resorting to the power equivalence $P_c = P_f$ the remaining quantity, either current or voltage, automatically follows. In the terminology that was introduced above, this means that the second step in the procedure, that is, the application of the circuit principle, selects one of the infinite number of *equivalent transmission lines* (with phase already equal for Z_f) to be the actual circuit transmission line representation of the microstrip transmission line. The selected Z_f value then becomes the characteristic circuit impedance Z_c. Depending on the contour choice introduced in the definition of either voltage or current, a different value of Z_c will be found.

By now it should be clear to the reader what the major lines of thought in constructing a circuit model for multiconductor lines are. Although the

1 GENERAL INTRODUCTION

above discussion was centred around the single waveguide case of a microstrip, the ideas can readily be extended to the multiconductor case. In the remaining part of this chapter we shall address some important additional questions regarding the circuit modelling problem.

1.2.6 *TEM waveguide modes and consistency with the classical circuit description*

The first question that might arise concerns the consistency of the circuit model that is obtained in the way discussed above with the classical circuit modelling described in Section 1.2.2. Indeed, as was argued above in discussing equations (1.1) and (1.2) for sufficiently low frequencies the flux contributions in (1.2) are negligible and voltage and current can be defined at the same time. In the second step of our circuit model selection procedure, however, we explicitly demanded that only one of these quantities should be defined by a path integral. We will show in Chapter 2 that applying the procedure of Section 1.2.5 is indeed compatible with the simultaneous definition of voltage and current at low frequencies. Although in general different ways of defining either V or I lead to different behaviour of the characteristic impedance as a function of frequency, all the dispersion curves for the impedances will be found to start from the same value at low frequencies.

Some waveguides, such as a coaxial cable (see Fig. 1.4) carry a TEM mode. This means that at all frequencies the longitudinal field components, that is, E_x (D_x) and H_x (B_x) in Fig. 1.4, of the considered mode are zero. As a consequence of this, and more particularly of the fact that the flux contributions from the right-hand sides of (1.2) vanish, it again becomes possible to define voltage and current simultaneously and for all frequencies. Hence, the derivation of an equivalent circuit transmission line model for such wave-

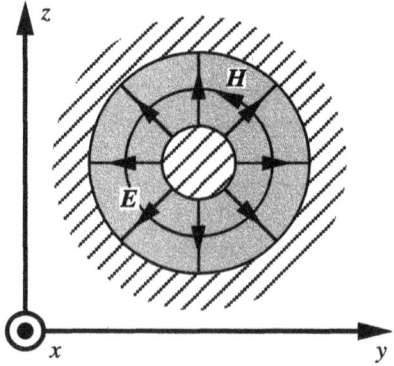

Fig. 1.4. Cross-section of a coaxial cable carrying a TEM mode.

1.2 CIRCUIT MODELLING OF MULTICONDUCTOR LINES

guide modes is simpler and more straightforward than for a general hybrid waveguide mode. With a tutorial purpose in mind, the subject matter of Chapter 2 is organized in such a way that the principles discussed in Section 1.2.5 are first explained for a TEM mode and are then followed by the more general treatment for a hybrid mode. This is no longer the case in Chapters 3 and 4 where we immediately tackle the hybrid case.

1.2.7 *Power–current and power–voltage model*

The second question that arises concerns the choice of the path P in the definition of the voltage or its counterpart c in the definition of the current. The criterion which is often put forward in the literature in connection with this particular problem is that of the TEM-like character of the resulting characteristic impedance [14]. This suggests that one particular choice for P or c in (1.1) is to be preferred over another if the resulting impedance keeps its low frequency, that is, its TEM value over as large a frequency band as possible. Although the ultimate, and in fact sole, criterion for preferring a particular circuit model is its ability to predict the real behaviour of the modelled structure in its network environment, no systematic investigations have been carried out to assess if the TEM criterion discussed above satisfies this demand. However, it will become clear from the subsequent discussion that the physics of the problem certainly plays a paramount role in the selection of a voltage or a current definition and hence in the definition of a characteristic impedance.

For multiconductor transmission lines of the coupled (coplanar or non-planar) microstrip type it has been shown that selecting the conduction current through each conductor as the current of the equivalent circuit model leads to a TEM-like character of the resulting impedances. This means that the path used to define the current in equation (1.1) is given by c_{con} for the microstrip line of Fig. 1.5(a) and that the same definition is adopted in the multiconductor case of Fig. 1.5(b). From a physical point of view it is quite clear that it is precisely this conduction current that will flow into the lumped circuit elements connected to the ports of the multiconductor line. From a mathematical point of view the reason for the TEM-like character of the resulting impedances resides in the fact that the presence of the air–dielectric interface on Fig. 1.5(a), or more generally speaking the presence of dielectric interfaces in Fig. 1.5(b), has a first-order effect on the electric field but only a second-order effect on the transversal magnetic field and the current flowing through each conductor [14]. The circuit model based on the above choice of the current together with the power equivalence principle is called the *power–current* model.

The above reasoning clearly indicates that the choice of a voltage or a current definition depends on the physics of the problem. The power–current

10   1 GENERAL INTRODUCTION

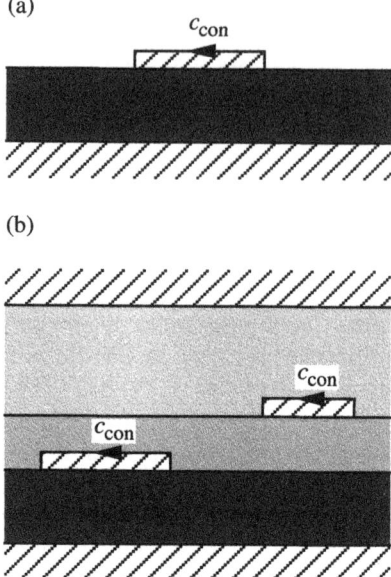

Fig. 1.5. Definition of the integration path for the current in the power–current formulation: (a) microstrip line; (b) coupled non-coplanar striplines.

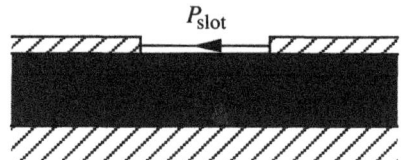

Fig. 1.6. Definition of the integration path for the voltage in the power–voltage formulation for the slotline.

model is certainly not applicable to every situation. In the particular case of the slotline [27] the voltage across the slot, that is, defined along the path P_{slot} in Fig. 1.6, is a much more natural choice as a circuit quantity. The circuit model based on this particular choice is called the *power–voltage* model.

In Chapter 2 we have chosen to illustrate both approaches, the power–current and the power–voltage models, on the microstrip line. The comparison of both approaches on the same structure is not only instructive but also allows us to confront both ways of obtaining a circuit model with the so-called current–voltage approach in which both current and voltage are defined at the same time while violating the power equivalence principle. At the end of Chapter 2 different models are discussed for several types of

1.2 CIRCUIT MODELLING OF MULTICONDUCTOR LINES 11

waveguide structures. In Chapter 3 the power–current formulation is generalized to the two coupled lossless waveguide case. In this case a pair of circuit voltages and currents has to be introduced. They are related in a linear way to the pair of field voltages and currents of the modal field description of the two lowest-order modes of the waveguide. This relation is again determined by applying the power equivalence and circuit principles.

Chapter 4 shows how two, in principle arbitrary, linear transformations, one between field currents and circuit currents and one between field voltages and circuit voltages, can be invoked to build an equivalent circuit model for an arbitrary number of coupled lossy conductors. Furthermore it is proven that the power equivalence principle leads to a relationship between both linear transformations and it is shown how the power–current or the power–voltage model can be introduced.

At the beginning of Section 1.2.4 it was mentioned that it is generally accepted that circuit modelling should be based on full-wave analysis. In this book we start from a modal description of the waveguides under study. Our circuit modelling is explicitly based on the fundamental modes propagated by the multiconductor waveguide. Even if we start from the assumption that no higher-order modes are propagated at the considered frequencies, these modes will contribute to the reactive power exchange at the ports of the waveguide. This problem deserves a more exhaustive study; some attention is devoted to it in Chapter 2.

1.2.8 *Reciprocity of the circuit model*

If no non-reciprocal materials are present in the multiconductor waveguide structure it would only be natural to expect the equivalent circuit model to be reciprocal as well. Although this aspect of the problem was actually never explicitly taken into account in the above discussion, it might perhaps still come as a surprise that the circuit models obtained are not reciprocal. In Chapter 4 this particular problem is discussed in detail for the general lossy multiconductor waveguide case. A general condition is derived that must be satisfied by the two linear transformations discussed at the end of Section 1.2.7 for the circuit model to be reciprocal. This condition shows that reciprocity is automatically guaranteed for lossless structures as a consequence of the power orthogonality of the modes. The problem of reciprocity was also discussed in [17]. This paper presents curves giving a measure of the non-reciprocity of the power–current model in the case of lossless structures. Due to the fact that the calculations for the circuit model in that paper are based on a somewhat *ad hoc* division of the total propagated power over the different conductors [12], the reciprocity turns out to be not guaranteed in that case either. Chapter 3 presents some details on this power division in the two coupled lossless waveguide case. A more extensive treatment for the multiconductor case can be found in [26].

1.2.9 Characteristic impedance matrix, impedance and admittance matrix, line-mode impedances

In circuit theory a single transmission line can be described either in terms of its complex propagation constant β and its complex characteristic impedance Z_c, or by its impedance z_c and admittance y_c per unit of length. These parameters are related to each other in a very simple way: $\beta^2 = -z_c y_c$ and $Z_c = \sqrt{z_c/y_c}$. For a single-conductor waveguide such as a coaxial cable the fundamental mode is a TEM mode and, assuming that no losses are present, $z_c = j\omega L$ and $y_c = j\omega C$ (for an $e^{j\omega t}$ time dependence) where L and C are the capacitance and inductance per unit of length. If L and C are frequency independent the propagation constant of the TEM mode is proportional to the frequency, that is, there is no signal dispersion (at least for that particular mode). For a general hybrid mode, such as for the microstrip line, the above description is still valid but all parameters become frequency dependent. For the single waveguide case, the representation of the circuit model in terms of a frequency-dependent L and C value is discussed in Chapter 2. To complete the picture, a frequency-dependent resistance R and conductance G (per unit of length) are introduced.

In Chapter 3, the above notions regarding the circuit description of a single waveguide are extended to the two coupled lossless waveguide structure. This involves the introduction of the propagation constant of the two fundamental modes, of a 2 × 2 impedance and admittance matrix, and of a set of impedances which are known as the line-mode impedances [3], [26]. It is very important not to confuse these line-mode impedances with the generalization of the characteristic impedance Z_c to a characteristic impedance matrix [28]. Once the circuit model for the coupled lines has been determined it is possible to express the circuit voltage and the circuit current associated with each conductor as the sum of contributions coming from each of the two modes. The impedances defined by these partial voltages and currents of each mode are the line-mode impedances. For two lines four such line-mode impedances can be defined.

It is also possible to extend the characteristic impedance concept to the multiconductor waveguide case. For a single line this characteristic impedance describes the ratio of circuit voltage to circuit current for a wave propagating down an infinitely long line. The second part of Chapter 4 presents the generalization of the characteristic impedance concept to multiconductor lines. The starting point of the discussion is the generalized telegrapher's equations [29]. Starting from this very general representation of a set of coupled circuit transmission lines we not only present the extension of the impedance and admittance matrix to the multiconductor case but also rigorously define a propagation matrix for the circuit voltages and a propagation matrix for the circuit currents. Attention is also devoted to the definition of line-mode impedances. Finally, the *R-L-C-G* description for the single waveguide case is extended to the multiconductor case and the resulting *R*, *L*, *C*, and *G* matrices are expressed in terms of the modal fields [26]. This *R-L-C-G* description is particularly suited to the physical interpretation of the circuit model parameters as it extends the well-known and physically well-understood quasi-TEM description of multiconductor lines [13], [20] to higher frequencies.

References

[1] Pitzalis, O. (1989). Microwave to mm-wave CAE: concept to production. *Microwave Journal*. 1989 State of the Art Reference, 15–47.

[2] Jansen, R. H. and Wiemer, L. (1984). Multiconductor hybrid-mode approach for the design of MIC couplers and lumped elements including loss, dispersion and parasitics. In *Proceedings of the 14th European Microwave Conference*, pp. 430–435, Liège.

[3] Chang, F.-Y. (1970). Transient analysis of lossless coupled transmission lines in a nonhomogeneous dielectric medium. *IEEE Transactions on Microwave Theory and Techniques*, **MTT-18**, 616–626.

[4] Fukuoka, Y., Zhang, Q., Neikirk, D. P., and Itoh, T. (1985). Analysis of multilayer interconnection lines for a high-speed digital integrated circuit. *IEEE Transactions on Microwave Theory and Techniques*, **MTT-33**, 527–532.

[5] Djordjevic, A. R., Sarkar, T. K., and Harrington, R. F (1987). Time-domain response of multiconductor transmission lines. *Proceedings of the IEEE*, **75**, 743–764.

[6] Carin, L., Xu, Q., Webb, K. J., and McClintock, J. A. (1987). Analysis of VLSI interconnect structures. In *1987 IEEE MTT-S International Microwave Symposium Digest*, pp. 625–628, Las Vegas.

[7] Dhaene, T. and De Zutter, D. (1992). Extended scattering matrix approach for transient analysis of coupled dispersive lossy transmission lines with arbitrary Loads. *Electromagnetics* (in press).

[8] Dhaene, T. and De Zutter, D. (1992). Time domain analysis of uniformly coupled lossy transmission lines with arbitrary loads. In *Proceedings of the 1992 International Symposium on Circuits and Systems*, pp. 2402–2404, San Diego.

[9] Olyslager, F., Dhaene, T., and De Zutter, D. (1992). Time and frequency domain study of the propagation in lossy multilayered waveguide and antenna structures based on a rigorous full-wave analysis. In *Proceedings of the 1992 IEEE-APS International Symposium*, pp. 1504–1507, Chicago.

[10] Marx, K. D. (1973). Propagation modes, equivalent circuits and characteristic terminations for multiconductor transmission lines with inhomogeneous dielectrics. *IEEE Transactions on Microwave Theory and Techniques*, **MTT-21**, 7, 450–457.

[11] Nagai, N. (1977). Modal decomposition and equivalent circuit representation of lossless coupled multiwire lines. In *Equivalent Circuit Representation of Coupled Multiwire Lines and its Application.* (ed. N. Nagai and R. Matori). Monograph Series, Rest. Inst. Appl. Electricity, Sapporo, Japan, No. 24, S. 7–30.

[12] Jansen, R. H. (1979). Unified user-oriented computation of shielded, covered and open planar microwave and millimeter-wave transmission-line characteristics. *IEE Journal of Microwaves, Optics and Acoustics*, **3**, 14–22.

[13] Lindell, I. V. (1981). On the quasi-TEM modes in inhomogeneous multiconductor transmission lines. *IEEE Transactions on Microwave Theory and Techniques*, **MTT-29**, 8, 812–817.

[14] Jansen, R. H. and Kirschning, M. (1983). Arguments and an accurate model for the power-current formulation of microstrip characteristic impedance. *Arch. Elek. Übertragung*, **37**, 3/4, 108–112.

[15] Getsinger, W. J. (1983). Measurement and modeling of the apparent characteristic impedance of microstrip. *IEEE Transactions on Microwave Theory and Techniques*, **MTT-31**, 8, 624–632.

[16] Jansen, R. H. (1985). The spectral domain approach for microwave integrated circuits. *IEEE Transactions on Microwave Theory and Techniques*, **MTT-33**, 10, 1043–1056.

[17] Wiemer, L. and Jansen, R. H. (1986). Reciprocity related definition of strip characteristic impedance for multiconductor hybrid-mode transmission lines. *Microwave and Optical Technology Letters*, **1**, 22–25.

[18] Brews, J. R. (1986). Transmission line models for lossy waveguide interconnections in VLSI. *IEEE Transactions on Electronic Devices*, **ED-33**, 9, 1356–1365.

[19] Brews, J. R. (1987). Characteristic impedance of microstrip lines. *IEEE Transactions on Microwave Theory and Techniques*, **MTT-35**, 1, 30–34.

[20] Lindell, I. V. and Gu, Q. (1987). Theory of time-domain quasi-TEM modes in inhomogeneous multiconductor transmission lines. *IEEE Transactions on Microwave Theory and Techniques*, **MTT-35**, 10, 893–897.

[21] Tripathi, V. K. and Lee, H. (1989). Spectral-domain computation of characteristic impedances and multiport parameters of multiple coupled microstrip lines. *IEEE Transactions on Microwave Theory and Techniques*, **MTT-37**, 1, 215–221.

[22] Faché, N. and De Zutter, D. (1989). Circuit parameters for single and coupled microstrip lines by rigorous full-wave space-domain analysis. *IEEE Transactions on Microwave Theory and Techniques*, **MTT-37**, 2, 421–425.

[23] Carin, L. and Webb, K. J. (1989). Characteristic impedance of multilevel, multiconductor hybrid mode microstrip. *IEEE Transactions on Magnetics*, **4**, 2947–2949.

[24] Faché, N. and De Zutter, D. (1990). New high-frequency circuit model for coupled lossless and lossy waveguide structures. *IEEE Transactions on Microwave Theory and Techniques*, **MTT-38**, 3, 252–259.

[25] Geshiro, M., Yagi, S., and Sawa, S. (1991). Analysis of slotlines and microstrip lines on anisotropic substrates. *IEEE Transactions on Microwave Theory and Techniques*, **MTT-39**, 1, 64–69.

[26] Dhaene, T. and De Zutter, D. (1992). CAD-oriented, general circuit description of uniform coupled lossy dispersive waveguide structures. *IEEE Transactions on Microwave Theory and Techniques*, **MTT-40**, 7, 1545–1554.

[27] Cohn, S. B. (1969). Slot line on a dielectric substrate. *IEEE Transactions on Microwave Theory and Techniques*, **MTT-17**, 10, 768–778.

REFERENCES

[28] Dhaene, T., De Zutter, D., and Criel, S. Analysis and modelling of coupled dispersive interconnection lines. *IEEE Transactions on Microwave Theory and Techniques*, (in press).
[29] Schelkunoff, S. A. (1955). Conversion of Maxwell's equations into generalized telegraphists' equations. *Bell System Technical Journal*, **34**, 995–1043.

2
CIRCUIT DESCRIPTION OF SINGLE WAVEGUIDE STRUCTURES

2.1 Introduction

The goal of this chapter is to derive the circuit model of an arbitrary single waveguide structure starting from the Maxwell equations which govern the electromagnetic behaviour of the structure. The circuit model is associated with the fundamental mode of the waveguide structure. The model must be extended in the case when the waveguide structure behaves as a multimode waveguide. However, this extension is not addressed in this book.

In the present chapter we distinguish TEM and hybrid modes. Although the circuit model for a waveguide structure supporting a TEM mode is well known and can be found in many standard textbooks on electromagnetic fields, it will be derived in this chapter. We will use the equivalent transmission line and circuit principles introduced in Chapter 1. These new principles are best illustrated by means of a TEM structure of which the circuit model is well known and unambiguous. Furthermore, a simultaneous treatment of the TEM and hybrid waveguides also makes it possible to discuss the differences between the derivation of the circuit models for the TEM and hybrid waveguides, in particular the impedance definition for hybrid waveguides.

The different steps carried out in this chapter to arrive at the circuit model starting from the electromagnetic field representation of the fundamental mode have already been explained in Chapter 1. The derivation of the circuit model of a TEM waveguide and of a hybrid waveguide is explained in parallel.

In Section 2.2 the modal field representation of a TEM and a hybrid mode is found as a solution to the sourceless Maxwell equations. The longitudinal dependence of the field components is described by means of a field voltage and current. This voltage–current pair is a solution to the classical transmission line equations and is associated with a field transmission line. This field transmission line has the same propagation constant as the mode in the waveguide. The field transmission lines differ from each other in the value of the characteristic impedance.

In Section 2.3 we start from the field transmission lines to derive equivalent transmission lines by applying the power equivalence principle. An equivalent transmission line propagates the same power with the same phase velocity and attenuation as the fundamental mode in the waveguide. The equivalent transmission lines differ from each other in the amplitude of the characteristic impedance. The phase of the characteristic impedance and the propagation constant are the same for all the transmission lines.

In Section 2.4 we construct the circuit model by selecting one of the equivalent transmission lines as the circuit transmission line model for the waveguide. To that end we will use the circuit principle to find the amplitude of the characteristic impedance. This can be done in a unique way for TEM waveguides, but no unique definition of the characteristic impedance exists for a hybrid waveguide. This is a consequence of the existence of longitudinal electromagnetic field components which make the line voltage and current dependent upon the path or contour. As was already mentioned in Chapter 1 we will introduce three impedance definitions in this chapter: a power–voltage, a power–current, and a voltage–current definition. We will evaluate the applicability of these definitions for different types of microwave structures and show the difference between the definitions for a microstrip.

2.2 Modal field representation

2.2.1 *TEM mode*

The study of a waveguide which propagates a single TEM mode will be exemplified by the coaxial line. The cross-section of the line is sketched in Fig. 2.1. The structure consists of an inner and an outer conductor, which are both perfectly conducting. The medium between the conductors is homogeneous and isotropic. It is characterized by its permittivity ε and permeability μ. For lossless structures ε and μ are real constants. In this

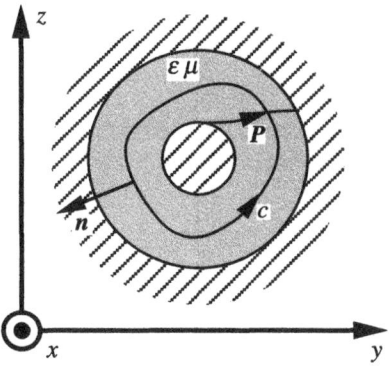

Fig. 2.1. Cross-section of a coaxial line.

chapter we allow electric and magnetic losses, in which case ε and μ take complex values.

In the general representation of a TEM mode the electric and magnetic fields are written as the product of a scalar function which depends only on the longitudinal coordinate x and a transversal vector which depends only on the transversal coordinates y and z:

$$\mathbf{E}(x, y, z) = V_f(x)\mathbf{E}_t(y, z)$$
$$\mathbf{H}(x, y, z) = I_f(x)\mathbf{H}_t(y, z).$$
(2.1)

The scalar functions $V_f(x)$ and $I_f(x)$ are called the field voltage and the field current. The reason for choosing these particular names will become clear shortly. The electric field $\mathbf{E}_t(y, z)$ and the magnetic field $\mathbf{H}_t(y, z)$ describe the field pattern in the cross-section of the waveguide. The subscript 't' refers to the transversal coordinates y and z. The field voltage V_f and the field current I_f have dimensions volt and ampere, respectively. The transversal fields $\mathbf{E}_t(y, z)$ and $\mathbf{H}_t(y, z)$ have dimension m^{-1}.

In order to find the longitudinal functions $V_f(x)$ and $I_f(x)$ and the fields $\mathbf{E}_t(y, z)$ and $\mathbf{H}_t(y, z)$ for the structure of Fig. 2.1 we substitute (2.1) into Maxwell's equations and apply the boundary conditions of the electromagnetic field at the inner and outer conductor.

Substitution of (2.1) in the curl equations of Maxwell and separating the transversal and longitudinal contributions results in the following set of equations:

$$\nabla_t \times \mathbf{E}_t(y, z) = \mathbf{0}$$

$$\nabla_t \times \mathbf{H}_t(y, z) = \mathbf{0}$$

$$\frac{dV_f(x)}{dx}[\mathbf{1}_x \times \mathbf{E}_t(y, z)] = -j\omega\mu I_f(x)\mathbf{H}_t(y, z)$$
(2.2)

$$\frac{dI_f(x)}{dx}[\mathbf{1}_x \times \mathbf{H}_t(y, z)] = j\omega\varepsilon V_f(x)\mathbf{E}_t(y, z).$$

The vector $\mathbf{1}_x$ is the unit vector in the x-direction. The operator ∇_t operates only on the transversal coordinates y and z and is defined by the following expression:

$$\nabla_t = \frac{\partial}{\partial y}\mathbf{1}_y + \frac{\partial}{\partial z}\mathbf{1}_z$$
(2.3)

in which $\mathbf{1}_y$ and $\mathbf{1}_z$ represent the unit vectors in the y- and z-directions respectively. From the first two equations of (2.2) it follows that $\mathbf{E}_t(y, z)$ and

2.2 MODAL FIELD REPRESENTATION

$\mathbf{H}_t(y, z)$ can each be written as the gradient of a potential function:

$$\begin{aligned} \mathbf{E}_t(y, z) &= -\nabla_t \phi(y, z) \\ \mathbf{H}_t(y, z) &= -\nabla_t \psi(y, z). \end{aligned} \quad (2.4)$$

Substitution of (2.1) in the divergence equations of Maxwell yields two scalar equations for $\mathbf{E}_t$ and $\mathbf{H}_t$:

$$\begin{aligned} \nabla_t \cdot \mathbf{E}_t(y, z) &= 0 \\ \nabla_t \cdot \mathbf{H}_t(y, z) &= 0. \end{aligned} \quad (2.5)$$

Combination of (2.4) with the divergence equations (2.5) shows that the potentials ϕ and ψ are a solution to the Laplace equations:

$$\begin{aligned} \nabla_t^2 \phi &= 0 \\ \nabla_t^2 \psi &= 0. \end{aligned} \quad (2.6)$$

However, the two potentials ϕ and ψ are not independent. In order to find the relation between these potentials we substitute the gradient form (2.4) of the electric and magnetic field into the last two equations of (2.2):

$$\begin{aligned} \frac{dV_f}{dx}(\mathbf{1}_x \times \nabla_t \phi) &= -j\omega\mu I_f \nabla_t \psi \\ \frac{dI_f}{dx}(\mathbf{1}_x \times \nabla_t \psi) &= j\omega\varepsilon V_f \nabla_t \phi. \end{aligned} \quad (2.7)$$

As a next step we separate the longitudinal and the transversal dependencies in (2.7). This yields a relationship between the two potentials ϕ and ψ:

$$\nabla_t \psi = \chi(\mathbf{1}_x \times \nabla_t \phi). \quad (2.8)$$

The constant of separation of variables χ is an arbitrary complex proportionality factor. As a by-product we have also found two differential relations between the field voltage and the field current. Combination of (2.7) and (2.8) indeed results in two equations relating the longitudinal functions $V_f(x)$ and $I_f(x)$:

$$\begin{aligned} \frac{dV_f(x)}{dx} + j\chi\omega\mu I_f(x) &= 0 \\ \frac{dI_f(x)}{dx} + j\frac{\omega\varepsilon}{\chi} V_f(x) &= 0. \end{aligned} \quad (2.9)$$

$V_f(x)$ and $I_f(x)$ are solutions to the transmission line equations for the voltage and the current on a transmission line. For this reason $V_f(x)$ and $I_f(x)$ are called the field voltage and field current of the mode. The general form of the classical transmission line equations for the voltage and current on a transmission line is given by

$$\frac{dV_f(x)}{dx} + j\beta_f Z_f I_f(x) = 0$$

$$\frac{dI_f(x)}{dx} + \frac{j\beta_f}{Z_f} V_f(x) = 0. \tag{2.10}$$

The equations (2.9) can be cast in this form by choosing the complex characteristic impedance Z_f and the complex propagation constant β_f of the transmission line to satisfy

$$Z_f = \chi \sqrt{\frac{\mu}{\varepsilon}} \tag{2.11}$$

and

$$\beta_f = \omega \sqrt{\varepsilon \mu}. \tag{2.12}$$

For the transmission line associated with V_f and I_f we use the term *field transmission line*. The associated complex impedance Z_f and the complex propagation constant β_f are called the field impedance and field propagation constant. The general solution to (2.10) consists of two waves travelling in opposite directions:

$$V_f(x) = A\,e^{-j\beta_f x} + B\,e^{j\beta_f x}$$

$$I_f(x) = \frac{1}{Z_f}(A\,e^{-j\beta_f x} - B\,e^{j\beta_f x}). \tag{2.13}$$

The constants A and B are determined by the boundary conditions at the beginning and the end of the waveguide. The propagation constant β_f is fixed and depends only on the material constants of the medium between the conductors. The characteristic impedance Z_f is proportional to χ and is therefore still an arbitrary complex number. Each choice of χ corresponds with a different field transmission line representation (2.13). All the field transmission lines propagate waves with the same propagation constant but with a different impedance.

From now on, the common propagation constant is simply denoted by β and is called the propagation constant of the mode. In general, the propagation constant β has a real and an imaginary part:

$$\beta = \beta' - j\beta''. \tag{2.14}$$

In the case of a lossless structure the attenuation constant β'' equals zero and the propagation constant is real. In the following we will select the

2.2 MODAL FIELD REPRESENTATION

square root in (2.12) in such a way that the real part of β is positive. This automatically leads to a non-negative attenuation constant.

The boundary conditions on the perfectly conducting inner and outer conductor require that the tangential components of the electric field be zero. Thus, the potential ϕ is a constant function on the conductors and takes the value ϕ_0 on the outer conductor and ϕ_1 on the inner conductor. The potential difference between the two conductors is denoted by $\Delta\phi = \phi_1 - \phi_0$.

To summarize, as a result of the substitution of (2.1) into Maxwell's equations and the application of the boundary conditions the following modal field representation is derived:

$$\mathbf{E}(x, y, z) = -(A\,e^{-j\beta x} + B\,e^{j\beta x})\nabla_t \phi(y, z)$$

$$\mathbf{H}(x, y, z) = -\frac{1}{Z_f}(A\,e^{-j\beta x} - B\,e^{j\beta x})\chi[\mathbf{1}_x \times \nabla_t \phi(y, z)]. \tag{2.15}$$

The total electromagnetic field can be expressed by a scalar electric potential ϕ which is a solution to the Laplace equation with Dirichlet boundary conditions on the inner and outer conductors:

$$\nabla_t^2 \phi = 0$$
$$\phi = \phi_1 \quad \text{on the inner conductor}$$
$$\phi = \phi_0 \quad \text{on the outer conductor} \tag{2.16}$$
$$\Delta\phi = \phi_1 - \phi_0.$$

Arbitrary constants can be placed in front of the potential ϕ but we will assume that ϕ is a real function.

The general representation (2.15) of the TEM mode shows that the electric and magnetic fields propagating in the positive x-direction are perpendicular:

$$\mathbf{1}_x \times \mathbf{E}^+ = Z_0 \mathbf{H}^+. \tag{2.17}$$

The proportionality factor Z_0 is the familiar wave impedance which is given by:

$$Z_0 = \sqrt{\frac{\mu}{\varepsilon}}. \tag{2.18}$$

We note that Z_0 has a unique value which depends on the characteristics of the medium between the conductors and is independent of the shape of the conductors. The superscript '+' in (2.17) emphasizes the fact that the fields belong to a wave propagating in the positive x-direction. For fields propagating in the negative x-direction Z_0 must be replaced by $-Z_0$ in (2.17).

We have shown that any TEM mode can be represented by transversal eigenfields $\mathbf{E}_t$ and $\mathbf{H}_t$ which are x-independent and a field–voltage–field-current pair (V_f, I_f) which behaves as the voltage and current pair along an x-directed field transmission line.

The two degrees of freedom in this general representation are the potential difference $\Delta\phi$ between the inner and outer conductors and the complex constant χ. The latter however does not appear in the total field representation as it cancels out in the expression for $\mathbf{H}$ because both $\mathbf{H}_t$ and Z_f are proportional to χ. χ is used only in the determination of I_f and $\mathbf{H}_t$ individually. The wave coefficients A and B depend upon χ, $\Delta\phi$, and the circuit in which the waveguide is embedded.

In Section 2.3, the application of the power equivalence principle and the circuit principle will determine the values of χ and $\Delta\phi$. We refer the reader to Chapter 1 for a definition of these principles.

2.2.2 Hybrid mode

A typical waveguide which propagates a hybrid mode is the microstrip depicted in Fig. 2.2. The waveguide consists of two homogeneous media: a

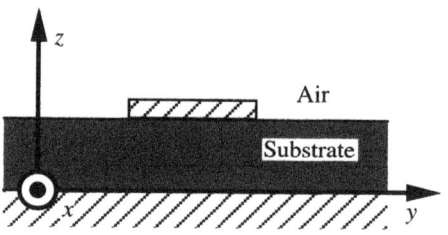

Fig. 2.2. Geometry of a microstrip structure.

substrate on top of a ground plane and a semi-infinite top layer. The strip is located at the interface between the two layers.

The electric and magnetic field associated with a hybrid mode consists of not only a transversal but also a longitudinal field component. Each component can be written as the product of a scalar function depending on the longitudinal coordinate and a vector depending on the transversal coordinates. Hence, the electromagnetic field takes the following general form:

$$\mathbf{E}(x, y, z) = A(x)\mathbf{E}_t(y, z) + B(x)\mathbf{E}_l(y, z)$$
$$\mathbf{H}(x, y, z) = C(x)\mathbf{H}_t(y, z) + D(x)\mathbf{H}_l(y, z). \qquad (2.19)$$

The scalar functions $A(x)$, $B(x)$, $C(x)$, and $D(x)$ and the fields $\mathbf{E}_t(y, z)$, $\mathbf{E}_l(y, z)$, $\mathbf{H}_t(y, z)$, and $\mathbf{H}_l(y, z)$ are still unknown. In order to find these unknowns we

2.2 MODAL FIELD REPRESENTATION

substitute the general field representation (2.19) into Maxwell's equations and the associated boundary conditions.

The Maxwell curl equations yield the following four vector equations for the modal field (2.19):

$$A(x)(\nabla_t \times \mathbf{E}_t) + j\omega\mu(y, z)D(x)\mathbf{H}_l = 0$$

$$\frac{dA(x)}{dx}(\mathbf{1}_x \times \mathbf{E}_t) + B(x)(\nabla_t \times \mathbf{E}_l) + j\omega\mu(y, z)C(x)\mathbf{H}_t = 0 \quad (2.20)$$

$$C(x)(\nabla_t \times \mathbf{H}_t) - j\omega\varepsilon(y, z)B(x)\mathbf{E}_l = 0$$

$$\frac{dC(x)}{dx}(\mathbf{1}_x + \mathbf{H}_t) + D(x)(\nabla_t \times \mathbf{H}_t) - j\omega\varepsilon(y, z)A(x)\mathbf{E}_t = 0$$

while the divergence equations result in the following two scalar equations:

$$A(x)\nabla_t \cdot [\varepsilon(y, z)\mathbf{E}_t] + \frac{dB(x)}{dx}\varepsilon(y, z)\mathbf{1}_x \cdot \mathbf{E}_l = 0 \quad (2.21)$$

$$C(x)\nabla_t \cdot [\mu(y, z)\mathbf{H}_t] + \frac{dD(x)}{dx}\mu(y, z)\mathbf{1}_x \cdot \mathbf{H}_l = 0.$$

In order to solve the equations (2.20) and (2.21) we separate the functions which depend on the longitudinal coordinate x from those which depend on the transversal coordinates y and z.

From the first equation of (2.20) it follows that $A(x)$ and $D(x)$ are proportional, while the third equation of (2.20) shows that $C(x)$ and $B(x)$ are proportional. The constants of separation of variables can be chosen freely. We have chosen the following relations:

$$R_0 D(x) = A(x)$$
$$R_0 C(x) = B(x) \quad (2.22)$$

where $R_0 = \sqrt{\mu_0/\varepsilon_0}$ is the characteristic impedance of free space. If we now assign the dimensions volt and ampere to the functions $A(x)$ and $C(x)$ then all the field components will have the same dimension m^{-1}. The reader might wonder why we make the particular choice (2.22) of an identical constant of separation of variables R_0 and why we do not introduce two arbitrary constants of separation of variables. We suggest that the reader goes along with our reasoning and we will come back to this point later.

From the second and fourth equations of (2.20) follow the same relations between the functions which depend only on the longitudinal coordinate as those that followed from the divergence equations (2.21):

$$\left.\frac{dA(x)}{dx}\right/C(x) = c_1$$
$$\left.\frac{dC(x)}{dx}\right/A(x) = c_2. \tag{2.23}$$

The separation constants c_1 and c_2 are arbitrary and independent of each other. Equations (2.23) can be cast in the form of the classical transmission line equations for the voltage and the current by expressing c_1 and c_2 in terms of the propagation constant $\beta = \beta_f$ and the characteristic impedance Z_f of the line:

$$c_1 = -j\beta Z_f$$
$$c_2 = \frac{-j\beta}{Z_f}. \tag{2.24}$$

At this point β and Z_f are arbitrary and independent constants, β has dimension m^{-1} and Z_f has dimension ohm. $A(x)$ plays the role of the voltage and $C(x)$ of the current on the transmission line. Based on this interpretation we introduce new names for $A(x)$ and $C(x)$:

$$A(x) = V_f(x)$$
$$C(x) = I_f(x). \tag{2.25}$$

$V_f(x)$ and $I_f(x)$ are called the field voltage and field current respectively, just as in the case of the TEM mode. The equations for field voltage and field current follow directly from equations (2.23) governing the x-dependence of the fields, the relation (2.24) between the pairs (c_1, c_2) and (β, Z_f), and the new nomenclature introduced in (2.25):

$$\frac{dV_f(x)}{dx} + j\beta Z_f I_f(x) = 0$$
$$\frac{dI_f(x)}{dx} + \frac{j\beta}{Z_f} V_f(x) = 0. \tag{2.26}$$

The general solution to equations (2.26) consists of two waves propagating in opposite directions:

$$V_f(x) = A\,e^{-j\beta x} + B\,e^{j\beta x}$$
$$I_f(x) = \frac{1}{Z_f}[A\,e^{-j\beta x} + B\,e^{j\beta x}] \tag{2.27}$$

where A and B are the constant amplitudes of the forward and backward

2.2 MODAL FIELD REPRESENTATION

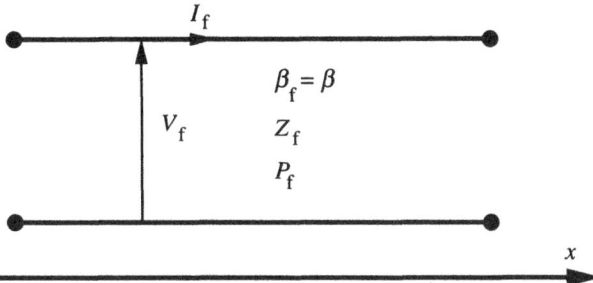

Fig. 2.3. Field transmission line with propagation constant $\beta_f = \beta$, impedance Z_f, and propagated power P_f.

propagating waves. They should not be confused with $A(x)$ and $B(x)$ introduced in (2.19). A field transmission line is depicted in Fig. 2.3. The power propagated in the field transmission line is denoted by P_f.

Combination of (2.19), (2.22), and (2.25) gives the new representation of the electromagnetic field in terms of the field voltage and field current:

$$\mathbf{E}(x, y, z) = V_f(x)\mathbf{E}_t(y, z) + R_0 I_f(x)\mathbf{E}_1(y, z) \qquad (2.28)$$
$$\mathbf{H}(x, y, z) = I_f(x)\mathbf{H}_t(y, z) + \frac{V_f(x)}{R_0}\mathbf{H}_1(y, z).$$

With the choices for the separation constants made in (2.22), (2.23), and (2.24), the curl equations governing the dependence of the fields on the transversal coordinates y and z can now be written in their final form:

$$(\nabla_t \times \mathbf{E}_t) + \frac{j\omega\mu(y, z)}{R_0}\mathbf{H}_1 = 0$$
$$j\beta Z_f(\mathbf{1}_x \times \mathbf{E}_t) - R_0 \nabla_t \times \mathbf{E}_1 - j\omega\mu(y, z)\mathbf{H}_t = 0 \qquad (2.29)$$
$$(\nabla_t \times \mathbf{H}_t) - j\omega\varepsilon(y, z)R_0\mathbf{E}_1 = 0$$
$$\frac{j\beta}{Z_f}(\mathbf{1}_x \times \mathbf{H}_t) - \frac{1}{R_0}\nabla_t \times \mathbf{H}_1 + j\omega\varepsilon(y, z)\mathbf{E}_t = 0.$$

The divergence equations take the following form:

$$\nabla_t \cdot [\varepsilon(y, z)\mathbf{E}_t] - \frac{j\beta R_0}{Z_f}\varepsilon(y, z)\mathbf{1}_x \cdot \mathbf{E}_1 = 0 \qquad (2.30)$$
$$\nabla_t \cdot [\mu(y, z)\mathbf{H}_t] - \frac{j\beta Z_f}{R_0}\mu(y, z)\mathbf{1}_x \cdot \mathbf{H}_1 = 0.$$

We now come back to our remark about the choice of identical constants of separation of variables in (2.22). The reader can check that if R_0 in the second equation of (2.22) were replaced by a different constant, say R'_0, then the only consequence of this replacement is that $\mathbf{E}_t R_0$ is replaced by $\mathbf{E}_t R'_0$. The additional degree of freedom is not relevant at all for the further development of our theory.

From equations (2.29) and (2.30) we can derive an eigenvalue equation for the propagation constant β. We distinguish an electric and a magnetic eigenvalue equation. The first one is an equation for the transversal electric field $\mathbf{E}_t$. The way to combine the equations in (2.29) and (2.30) to arrive at this equation is explained in [1]. We restrict ourselves to giving the final result:

$$\mu\left[\nabla_t \times \frac{1}{\mu}(\nabla_t \times \mathbf{E}_t)\right] - \nabla_t\left[\frac{1}{\varepsilon}\nabla_t \cdot (\varepsilon \mathbf{E}_t)\right] - (\beta^2 - \omega^2\varepsilon\mu)\mathbf{E}_t = \mathbf{0}. \quad (2.31)$$

The second one is a similar equation for the transversal magnetic field $\mathbf{H}_t$. The magnetic eigenvalue equation is [1]:

$$\varepsilon\left[\nabla_t \times \frac{1}{\varepsilon}(\nabla_t \times \mathbf{H}_t)\right] - \nabla_t\left[\frac{1}{\mu}\nabla_t \cdot (\mu \mathbf{H}_t)\right] - (\beta^2 - \omega^2\varepsilon\mu)\mathbf{H}_t = \mathbf{0}. \quad (2.32)$$

The equations (2.31) and (2.32) are called eigenvalue equations because solutions to (2.31) and (2.32) satisfying the boundary conditions exist only for specific values of the propagation constant β. These values may be discrete, continuous, or a combination of both, depending on the structure under evaluation. The eigenvalue equations do not depend on the choice of the characteristic impedance Z_f and therefore all the field transmission lines have the same propagation constant β which is also the propagation constant of the mode.

A solution to either (2.31) or (2.32) determines the field of a hybrid mode. However for a certain modal field the characteristic impedance Z_f is still arbitrary and one field component can always be multiplied by an arbitrary factor provided that the other field components and the field voltage and field current are also multiplied by suitable factors. These factors are determined by the curl and divergence equations (2.29) and (2.30). Removal of this non-uniqueness will be effected by the application of the power equivalence principle and of the circuit principle. These will be discussed in the next two sections.

2.3 Application of the power equivalence principle

2.3.1 *TEM mode*

A field transmission has the same propagation constant β as the mode in the waveguide. Consequently the field transmission line propagates its power

2.3 APPLICATION OF THE POWER EQUIVALENCE PRINCIPLE

with the same phase velocity and attenuation as the waveguide. However, in order to be an equivalent transmission line (using the terminology introduced in Chapter 1, Section 1.2.5) the field transmission line must also propagate the same amount of complex power as the waveguide. Imposing this requirement will yield the subclass of field transmission lines which are also equivalent transmission lines.

The complex power propagated by the field voltage and the field current in the field transmission line (2.13) is given by the following expression:

$$P(x) = \tfrac{1}{2} V_f(x) I_f^*(x). \tag{2.33}$$

For a mode propagating in the positive x-direction, we can use the relation $V_f = Z_f I_f$ between the field voltage and field current to produce equivalent expressions for $P(x)$. They are given by

$$P(x) = \frac{1}{2} \frac{|V_f(x)|^2}{Z_f^*} \tag{2.34}$$

and

$$P(x) = \tfrac{1}{2} Z_f |I_f(x)|^2 \tag{2.35}$$

where '| |' represents the absolute value and '*' the complex conjugate operator. From (2.34) and (2.35) we can see that in that particular case the phase of the power $P(x)$ and the phase of the characteristic impedance are the same.

On the other hand, the power propagated by the electromagnetic field in the waveguide follows from the integration of the longitudinal component of the Poynting vector over the cross-section of the waveguide. Simple mathematical manipulations lead to

$$P(x) = -\tfrac{1}{2} \chi^* V_f(x) I_f^*(x) (\Delta\phi) \oint_c \frac{\partial \phi}{\partial n} dc. \tag{2.36}$$

To obtain this expression we used the representation (2.15) of the modal field. The contour c encloses the inner conductor and is located between the inner and the outer conductor (Figure 2.1). **n** is perpendicular and external to the contour c.

Equalization of the power expressions (2.33) and (2.36) results in a relation between the potential difference $\Delta\phi$ and χ:

$$-\chi^* \Delta\phi \oint_c \frac{\partial \phi}{\partial n} dc = 1. \tag{2.37}$$

We note that the normal derivative of the potential in the integrand of the integral in (2.37) is proportional to the potential difference $\Delta\phi$. We still need an additional relation between χ and $\Delta\phi$ to determine both constants.

This relation will follow from the application of the circuit principle. However, from (2.37) we can already conclude that χ is not a complex but a real constant because the potential ϕ and its derivative are real. The factor χ appears in the expression (2.11) for the characteristic impedance of the field transmission line. For a general field transmission line χ is complex but for the equivalent transmission lines χ is real. For lossless structures the impedance of an equivalent transmission line will be an arbitrary real constant and from the power expressions (2.34) and (2.35) we can see that the phase of the power of a mode propagating in the positive x-direction is equal to zero. Observe that if $\Delta\phi > 0$ then $\partial\phi/\partial n < 0$ and vice versa.

2.3.2 Hybrid mode

The construction of an equivalent transmission line for a hybrid mode is handled in a similar way to that for a TEM mode. As a first step we calculate the power propagated in both the waveguide and in a field transmission line.

Substitution of the expression for the electromagnetic field (2.28) of the mode in the general expression for the power propagated through a waveguide gives the following result:

$$P(x) = \tfrac{1}{2} V_f(x) I_f^*(x) \int_S \int [\mathbf{E}_t(y,z) \times \mathbf{H}_t^*(y,z)] \cdot \mathbf{1}_x \, dy \, dz. \qquad (2.38)$$

The expression (2.38) contains only the transversal electric and magnetic fields. The longitudinal fields do not contribute to the power propagated in the longitudinal direction.

The power propagated in the field transmission line is given by the same expression (2.33) as for the TEM mode:

$$P(x) = \tfrac{1}{2} V_f(x) I_f^*(x). \qquad (2.39)$$

The equivalent power expressions (2.34) and (2.35) used in the TEM case also hold for a hybrid mode propagating in the positive x-direction.

As a next step we require that both power expressions (2.38) and (2.39) be equal. This results in a condition for the transversal electromagnetic field:

$$\int_S \int (E_y H_z^* - E_z H_y^*) \, dy \, dz = 1. \qquad (2.40)$$

The left-hand side of equation (2.40) is a complex number while the right-hand side is real. For this reason we are apt to think that the condition (2.40) cannot always be realized, or that there is not always an equivalent transmission line for the mode. However, by using the Maxwell equations the condition (2.40) can be transformed into a new condition for the transversal electric or transversal magnetic field. The calculations are

2.3 APPLICATION OF THE POWER EQUIVALENCE PRINCIPLE

presented in [1] so we restrict ourselves to giving the result:

$$\int_S \int \left[\frac{1}{\mu} (\nabla_t \times \mathbf{E}_t) \cdot (\nabla_t \times \mathbf{E}_t^*) - \omega^2 \varepsilon |\mathbf{E}_t|^2 \right] dy\, dz = \frac{-\omega \beta}{Z_f} \quad (2.41)$$

or

$$\int_S \int \left[\frac{1}{\varepsilon} (\nabla_t \times \mathbf{H}_t) \cdot (\nabla_t \times \mathbf{H}_t^*) - \omega^2 \mu |\mathbf{H}_t|^2 \right] dy\, dz = -\omega \beta Z_f. \quad (2.42)$$

The new expressions (2.41) and (2.42) are equivalent; they are related to each other through the Maxwell equations. Now we can show that these new conditions imply a restriction on the impedance, in particular on the phase of the impedance. The proof goes as follows for equation (2.41). For a given electromagnetic field [$\mathbf{E}, \mathbf{H}$] the transversal electric field $\mathbf{E}_t$ can always be multiplied by an arbitrary complex number δ provided that the other field components and field voltage and field current are adjusted accordingly so that the total electromagnetic field and the propagated power remain the same. The multiplication factors for the fields and the field quantities are shown in Table 2.1, which follows from the curl (2.29) and divergence (2.30) equations and the general field representation (2.28) of a mode. The derivation is left to the reader as an exercise.

Table 2.1 Multiplication factors for the fields and the field quantities

$\mathbf{E}_t \to \delta \mathbf{E}_t$	$\mathbf{H}_t \to \dfrac{\mathbf{H}_t}{\delta^*}$
$V_f \to \dfrac{V_f}{\delta}$	$I_f \to I_f \delta^*$
$\mathbf{H}_l \to \delta \mathbf{H}_l$	$\mathbf{E}_l \to \dfrac{\mathbf{E}_l}{\delta^*}$
$Z_f \to \dfrac{Z_f}{|\delta|^2}$	

Taking Table 2.1 into consideration, (2.41) and (2.42) fix only the phase of the impedance Z_f and not its amplitude. We obtained the same result as for the TEM mode: the equality of the power propagated in the waveguide and in the equivalent transmission line is always realizable. A mode has an infinite set of equivalent transmission lines which differ from each other in the amplitude of their characteristic impedance but which have the same impedance and power phase factor. For lossless structures this phase equals zero.

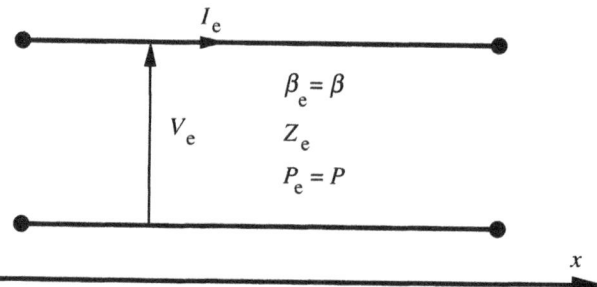

Fig. 2.4. Equivalent transmission line with propagation constant $\beta_e = \beta$, impedance Z_e, and propagated power $P_e = P$.

From now on, the voltage, current, and impedance of an equivalent field transmission line will be denoted by $V_e(x)$, $I_e(x)$, and Z_e respectively. $V_e(x)$ and $I_e(x)$ are solutions to the same equations (2.26) (or the identical equations (2.10) derived for the TEM case) as the ones for the field quantities. The general solution is given by

$$V_e(x) = A\,e^{-j\beta x} + B\,e^{j\beta x} = V_e^+(x) + V_e^-(x)$$
$$I_e(x) = \frac{1}{Z_e}(A\,e^{-j\beta x} - B\,e^{j\beta x}) = I_e^+(x) - I_e^-(x)$$
(2.43)

but now the equivalent field voltage and current have associated field components which satisfy the conditions (2.41) and (2.42). The propagation constant β_e of an equivalent transmission line remains unchanged and is the same as the propagation constant β of the mode. Hence we do not introduce a new notation and immediately replace β_e by β. The characteristic impedance Z_e is such that the equivalent transmission line propagates the same power as the waveguide. The equivalent transmission line is shown in Fig. 2.4.

In (2.43) we also redefined the total equivalent field voltage and current in terms of a contribution propagating in the positive x-direction (superscript '+') and one in the negative x-direction (superscript '−'). From the general representation (2.43) we can derive the impedance Z_e using one of the three definitions, as introduced by Schelkunoff [2]:

(1) the voltage–current definition:

$$Z_e = \frac{V_e^+(x)}{I_e^+(x)};$$
(2.44)

(2) the power–voltage definition:

$$Z_e = \frac{|V_e^+(x)|^2}{2[P^+(x)]^*};$$
(2.45)

2.3 APPLICATION OF THE POWER EQUIVALENCE PRINCIPLE

(3) the power–current definition:

$$Z_e = \frac{2P^+(x)}{|I_e^+(x)|^2}. \tag{2.46}$$

For each equivalent transmission line the three definitions result in the same impedance value because the three quantities V_e^+, I_e^+, and P^+ are related to each other through (2.39), that is, $P^+(x) = V_e^+ I_e^{+*}/2$. However, for another equivalent transmission line the impedance Z_e takes a different value.

The circuit parameters of an equivalent transmission line are the resistance R_e, the conductance G_e, the inductance L_e, and the capacitance C_e per unit length in the propagation direction. Note that R_e, G_e, L_e, and C_e are frequency dependent. These real quantities determine the propagation constant β and the characteristic impedance Z_e in the following way:

$$\beta^2 = -(R_e + j\omega L_e)(G_e + j\omega C_e) \tag{2.47}$$

and

$$Z_e^2 = \frac{R_e + j\omega L_e}{G_e + j\omega C_e}. \tag{2.48}$$

From the modal analysis of a hybrid waveguide follows the propagation constant β and the propagated power P^+ of a mode propagating in the positive x-direction. Only the phase of the power is of importance because a mode is a homogeneous solution to the sourceless Maxwell equations and all modal fields can always be multiplied by an arbitrary complex factor. To express R_e, G_e, L_e, and C_e in terms of β and the phase of P^+ we introduce the real and imaginary parts of β, P^+, and Z_e:

$$\beta = \beta' - j\beta'' \tag{2.49}$$

$$\begin{aligned} P^+ &= P'^+ + jP''^+ \\ Z_e &= Z_e' + jZ_e''. \end{aligned} \tag{2.50}$$

The constant β'' describes the attenuation of the mode in the longitudinal direction. The propagation is described by the real part β'. To eliminate the amplitude of the power we will work with the power quotient Q_P:

$$Q_P = \frac{P''^+}{P'^+} = \frac{Z_e''}{Z_e'}. \tag{2.51}$$

As we are considering a mode propagating in the positive direction (and not a combination of two modes propagating in opposite directions) the power quotient Q_P is independent of the cross-section in which we calculate the power P^+ because the real and imaginary parts have the same longitudinal dependence.

Using (2.47)–(2.49) and (2.51) we find the following relations between the circuit parameters and the parameters β and Q_P:

$$\frac{\omega L_e}{R_e} = \frac{\beta'' Q_P + \beta'}{\beta'' - \beta' Q_P} \qquad (2.52)$$

$$\frac{\omega C_e}{G_e} = \frac{\beta'' Q_P - \beta'}{\beta'' - \beta' Q_P} \qquad (2.53)$$

$$G_e R_e = \frac{\beta''^2 - \beta'^2 Q_P^2}{1 - Q_P^2} \qquad (2.54)$$

$$\omega^2 L_e C_e = \frac{\beta'^2 - \beta''^2 Q_P^2}{1 - Q_P^2}. \qquad (2.55)$$

The right-hand sides of (2.52)–(2.55) contain physical quantities which follow from the analysis of the hybrid mode and are therefore known. The left-hand sides contain the unknown circuit parameters. The four equations (2.52)–(2.55) are not independent; one equation can always be found from a suitable combination of the other three. For example, the first equation (2.52) can be obtained by dividing both sides of (2.55) by the product of (2.53) and (2.54). This dependence also means that the amplitude of the characteristic impedance can be chosen freely, a result we obtained earlier. This degree of freedom will be exploited in the application of the circuit principle.

2.4 Application of the circuit principle

2.4.1 Introduction

From the collection of equivalent transmission lines we will now select one line as the equivalent circuit model of the waveguide. This selection is based upon the circuit concept for the waveguide: the circuit or network model is this equivalent transmission line which most accurately describes the power exchange between the waveguide and its load and driver. This means that TEM concepts will have to be invoked by defining either a voltage or a current.

The circuit model of the waveguide can be interconnected with circuit models of the load and the driver as shown in Fig. 2.5. For the circuit quantities we use the subscript 'c'.

The propagation constant β of the circuit model in Fig. 2.5 is the same as the propagation constant of the mode. The characteristic impedance Z_c has a known phase factor because the circuit model is an equivalent

2.4 APPLICATION OF THE CIRCUIT PRINCIPLE

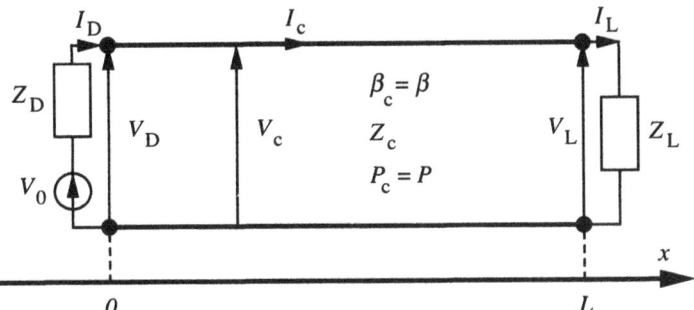

Fig. 2.5. Circuit model for a waveguide with a load and a driver.

transmission line in the sense discussed in Section 2.3. But the amplitude of Z_c has not yet been determined. This is the purpose of the present section.

We assume that the frequency is lower than the cut-off frequency of the first higher-order mode. Only one mode, the fundamental mode, is propagated in the waveguide. Once the frequency is higher than the cut-off frequency all the other propagating modes must be taken into account. How this is done is beyond the scope of the present discussion.

Although the waveguide propagates only one mode, there will be an infinite number of reflected modes at realistic load and driver interfaces. Except for the propagating mode all these modes decay exponentially and do not propagate power. In the case of losses one has to take into account the power dissipated by these modes. We will assume that for all practical purposes all the higher-order modes excited at one interface have died out at the other interface.

Before we derive the amplitude of the characteristic impedance Z_c we first review the circuit models for the load and the driver. We will also discuss the assumptions made in the construction of these models.

2.4.2 Circuit models for load and driver

The circuit models for the load and the driver follow from the solution of the quasi-static approximated Maxwell equations. In this approximation each circuit component can be described by means of a line voltage and a line current at its terminals. We will assume below that for the load the following relation between the line voltage and current applies:

$$V_L = Z_L I_L. \tag{2.56}$$

The subscript 'L' refers to the load and Z_L is the impedance of the load. We suppose that a similar expression applies for the driver:

$$V_D = V_0 + Z_D I_D \tag{2.57}$$

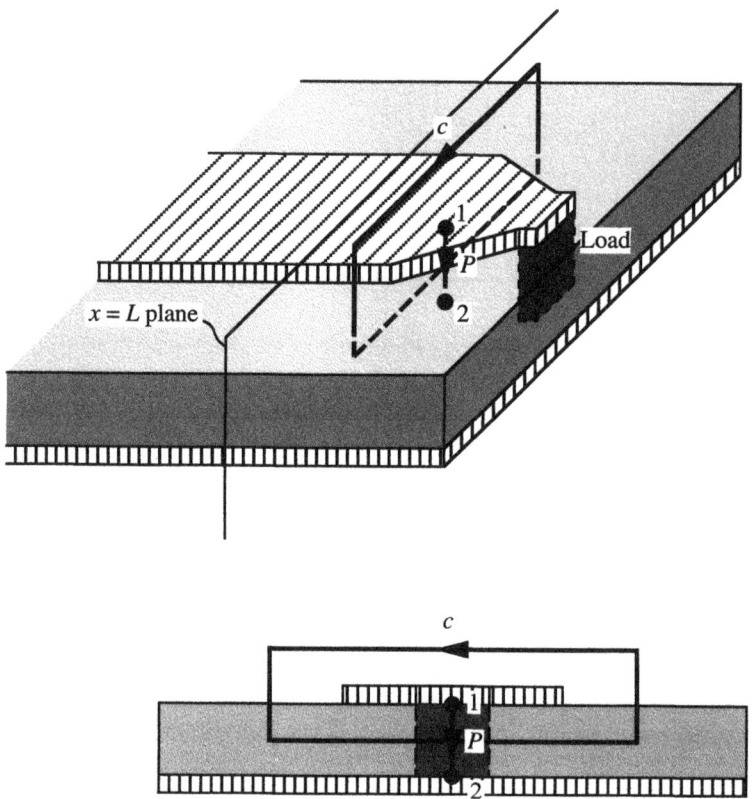

Fig. 2.6. Integration paths for the line voltage and line current of the load.

in which V_0 represents the driver voltage and Z_D the driver internal impedance. The subscript 'D' refers to the driver.

The integration paths for the line voltage and the line current of the load are sketched in Fig. 2.6. They are both located in the plane $x = L$. The line voltage is obtained by integrating the total transversal electric field along a path P from point 1 to 2:

$$V_L = \int_P \mathbf{E}_t^{tot} \cdot d\mathbf{l}. \tag{2.58}$$

The line current follows from the integration of the total transversal magnetic field along the closed contour c which encloses the conductor of the waveguide:

$$I_L = \oint_c \mathbf{H}_t^{tot} \cdot d\mathbf{l}. \tag{2.59}$$

2.4 APPLICATION OF THE CIRCUIT PRINCIPLE

Similar expressions to (2.58) and (2.59) hold for the line voltage and current at the driver interface. Further discussion is restricted to the load.

In the quasi-static approximation the line voltage V_L and the line current I_L are independent of the choice of the path P and the contour c, as discussed in Chapter 1. For example, if we choose another path P' from point 1 to point 2 then the value of V_L remains the same. In general, the difference between the integration of the field along path P and P' is proportional to the flux of the magnetic field through the surface S enclosed by both paths P and P' and the inner and outer conductor:

$$\int_P \mathbf{E}_t^{\text{tot}} \cdot d\mathbf{l} - \int_{P'} \mathbf{E}_t^{\text{tot}} \cdot d\mathbf{l} = -j\omega \int_S \int \mu \mathbf{H}_1^{\text{tot}} \cdot \mathbf{1}_x \, dy \, dz. \quad (2.60)$$

We assume that both conductors are perfectly conducting. In the quasi-static approximation the term on the right-hand side of (2.60) is neglected. This means that the time variation of the electromagnetic field is negligible compared to the space variations in the (y, z) plane. For a purely transversal electromagnetic field the equality between the two integrations along P and P' holds exactly. Similar conclusions can be drawn for the line current I_L.

In a general waveguide, the reflection of an incoming wave at the load (or driver) is accompanied by the excitation of exponentially decaying higher-order modes. The total field is not transversal even when the fundamental mode is a TEM mode. Indeed, a TEM mode is transversal but the higher-order modes excited at the load are not. The line voltage and current defined in (2.58) and (2.59) lose their unique values. In those cases we will restrict the integration paths P and c. We now discuss this restriction.

The line current in (2.59) consists of two contributions: the total longitudinal current flowing on the conductor and the total dielectric displacement current through the surface enclosed by the contour c. The latter can be neglected in the quasi-static approximation and equals zero for a TEM field. For those cases where the quasi-static approximation does not apply we will restrict the contour c to the circumference c_c of the conductor so that the line current equals the total longitudinal current flowing on the conductor. This is the power–current approach.

The path for the line voltage will be restricted to one specific path P in the plane $x = L$. Any other path results in another line voltage, the difference being a consequence of the flux of the magnetic field through the surface enclosed by the conductors, path P', and the load. This is the power–voltage approach. As stated in Chapter 1, voltage and current cannot be chosen freely at the same time.

The contour c_c selected in the power–current approach and the contour P selected in the power–voltage approach do not guarantee that the resulting circuit voltage and circuit current of the equivalent transmission line circuit model accurately describe the power exchange with the load. Indeed, the load must be such that the line voltage and the line current describe the power dissipated in the load. For the load, and again referring to Fig. 2.6, this power is given by

$$P_L = \tfrac{1}{2} V_L I_L^*. \tag{2.61}$$

If the dimensions of the load become too significant with respect to the wavelength, it will no longer be possible to describe this load in terms of line quantities and (2.61) will no longer apply. In that case the influence of the higher-order modes in the waveguide becomes important and a full three-dimensional field problem has to be solved at the load.

2.4.3 *Calculation of the amplitude of the characteristic impedance*

We will now apply the circuit principle discussed in Chapter 1 in order to calculate the amplitude of the characteristic impedance. We start from the voltage and current on an arbitrary equivalent field transmission line. At the load (driver) interface the field voltage and the field current must satisfy continuity relations with the line voltage over and the line current through the load (driver), and thus must also satisfy the circuit equations for the load (driver). The line voltage over and the line current through the load (driver) interface also follow a suitable integration of the total electric and total magnetic field respectively. These fields consist of the fundamental mode and a set of higher-order modes. From the continuity relations we will be able to define a characteristic impedance for the waveguide.

To find a value for the characteristic impedance which depends only on the waveguide and not on the load (driver) or the excited higher-order modes, approximations for the contributions from the higher-order modes are needed. For a TEM mode we will show that the contribution of the higher-order modes to the line current through and line voltage over the load (driver) can be neglected. For hybrid modes, however, things will be more complicated. For example, if we neglect the contribution of the higher-order modes to the line current then in general there will be a contribution of the higher-order modes to the line voltage. This contribution can be expressed in terms of the contribution of the fundamental mode to the line current using the circuit relation between the line voltage and the line current.

For TEM modes there will be three equivalent definitions of the impedance based on two out of the following three quantities: propagated power, line voltage, and line current, all three associated with the fundamental mode. The line current associated with the fundamental mode follows from the integration of the magnetic field of the fundamental mode along a closed contour c. The line voltage of the fundamental mode on the other hand is obtained by integrating the transversal electric field of the fundamental mode along a path P between the waveguide and the ground plane. For hybrid modes there will also be three equivalent definitions in each circuit model but they are based on three quantities of which only two can follow from the fundamental mode. The third one is based upon the fundamental mode and the higher-order modes. The circuit model will be named after the impedance definition which follows from two fundamental mode quantities. In the following we will perform all the calculations for the load. The conclusions are the same for the load as for the driver.

In order to make clear that we now select one of the equivalent transmission line representations to be the final circuit representation to be used for network simulation purposes, we will use the subscript 'c' for the associated voltage, current, and impedance. To recapitulate and to emphasize the different steps in our construction of a circuit model we remind the reader that we started out with the notation V_f, I_f, and Z_f (see (2.13) and (2.27)) indicating field quantities. We switched to V_e, I_e, and Z_e (see (2.43)) to indicate a selection of all possible field representations, that is, the ones termed equivalent in a power sense. We now further restrict the infinite set of equivalent lines to a single one based on the assumptions discussed in the present section and this final selection is denoted as V_c, I_c, and Z_c indicating that the sought-for circuit representation is obtained. Z_c is the circuit or characteristic impedance.

The circuit voltage and the circuit current propagating on the circuit transmission line are given by

$$V_c(x) = A\,e^{-j\beta x} + B\,e^{j\beta x}$$
$$I_c(x) = \frac{1}{Z_c}(A\,e^{-j\beta x} - B\,e^{j\beta x}). \quad (2.62)$$

At the interface of the load of Figure 2.6 they take the following values:

$$V_c(x = L) = V_L = A\,e^{-j\beta L} + B\,e^{j\beta L} \quad (2.63)$$

$$I_c(x = L) = I_L = \frac{1}{Z_c}(A\,e^{-j\beta L} - B\,e^{j\beta L}). \quad (2.64)$$

The voltage and current must satisfy the circuit relation (2.56) at the load interface. The voltage reflection coefficient

$$R_{V,L} = \frac{B}{A}e^{2j\beta L} \quad (2.65)$$

defined as the ratio of the amplitude of the reflected and incoming voltage wave at the load follows from the substitution of (2.65) in (2.63) and (2.64) and relationship (2.56):

$$R_{V,L} = \left(\frac{Z_c - Z_L}{Z_c + Z_L}\right) e^{2j\beta L}. \tag{2.66}$$

Once the amplitude of Z_c is determined $R_{V,L}$ can also be used to describe the power exchanged between the waveguide and the load.

The line voltage over and the line current through the load are also given by (2.58) and (2.59). The total field appearing in the integrands of (2.58) and (2.59) can be written as a combination of the fundamental mode and the higher-order modes excited by the discontinuity at the load interface. Under the assumptions that only the fundamental mode propagates in the waveguide and that the higher-order modes excited at one interface do not couple with the other interface, the transversal electric and magnetic fields at the load interface are given by

$$\mathbf{E}_t^{tot}(y, z) = A e^{-j\beta L}(1 + R_{V,L})\mathbf{E}_t(y, z) + \sum_{i=1}^{N} R_i A e^{j\beta_i L}\mathbf{E}_{t,i}(y, z) \tag{2.67}$$

and

$$\mathbf{H}_t^{tot}(y, z) = \frac{A e^{-j\beta L}}{Z_c}(1 - R_{V,L})\mathbf{H}_t(y, z) - \sum_{i=1}^{N} \frac{R_i A}{Z_i} e^{j\beta_i L}\mathbf{H}_{t,i}(y, z). \tag{2.68}$$

In a similar way as for the fundamental mode, a higher-order mode can be described by an equivalent transmission line with impedance Z_i and propagation constant β_i. The associated transversal field is $[\mathbf{E}_{t,i}, \mathbf{H}_{t,i}]$. There are only higher-order modes propagating in the negative x-direction. The (complex) amplitudes of these modes are described by a reflection coefficient R_i and the amplitude A of the incoming fundamental mode. From (2.67) and (2.68) we can calculate the line voltage (2.58) over and the line current (2.59) through the load:

$$V_L = A e^{-j\beta L}(1 + R_{V,L}) \int_P \mathbf{E}_t \cdot d\mathbf{l} + \sum_{i=1}^{N} R_i A e^{j\beta_i L} \int_P \mathbf{E}_{t,i} \cdot d\mathbf{l} \tag{2.69}$$

$$I_L = \frac{A e^{-j\beta L}}{Z_c}(1 - R_{V,L}) \oint_{c_c} \mathbf{H}_t \cdot d\mathbf{l} - \sum_{i=1}^{N} \frac{R_i A}{Z_i} e^{j\beta_i L} \oint_{c_c} \mathbf{H}_{t,i} \cdot d\mathbf{l}. \tag{2.70}$$

In the next step of our calculations we express the continuity of the line voltage and the line current at the load interface. For the voltage we have the equality of $V_c(x = L)$ (2.63) and V_L (2.69):

$$A e^{-j\beta L}(1 + R_{V,L}) = A e^{-j\beta L}(1 + R_{V,L}) \int_P \mathbf{E}_t \cdot d\mathbf{l} + \sum_{i=1}^{N} R_i A e^{j\beta_i L} \int_P \mathbf{E}_{t,i} \cdot d\mathbf{l}. \tag{2.71}$$

2.4 APPLICATION OF THE CIRCUIT PRINCIPLE

For the current we have the equality of $I_c(x = L)$ (2.64) and I_L (2.70):

$$\frac{A\,e^{-j\beta L}}{Z_c}(1 - R_{V,L}) = \frac{A\,e^{-j\beta L}}{Z_c}(1 - R_{V,L})\oint_{c_c} \mathbf{H}_t \cdot d\mathbf{l} - \sum_{i=1}^{N} \frac{R_i A}{Z_i} e^{j\beta_i L}\oint_{c_c} \mathbf{H}_{t,i} \cdot d\mathbf{l}. \tag{2.72}$$

To solve the equations (2.71) and (2.72) we note that the transversal electric or magnetic field of the fundamental mode can always be multiplied by an arbitrary complex factor δ. For example if we do this for the transversal electric field $\mathbf{E}_t(y, z)$ in (2.28) we can see from Table 2.1 that we must also divide the characteristic impedance Z_c by $|\delta|^2$, divide the transversal magnetic field $\mathbf{H}_t(y, z)$ by δ^*, divide the field voltage by δ and multiply the field current by δ^*. The left-hand side of equations (2.71) and (2.72) do not depend on the arbitrary choice of the multiplication factor δ for the fields of a mode, but the right-hand sides do, which means that the equality between the left- and right-hand sides can be fulfilled only for a particular value of δ. The solution of either equation fixes δ and all the field components of the mode and also the characteristic impedance (see (2.41) or (2.42)). The value of δ obtained from both equations will be the same but which contribution appears in the right-hand sides of (2.71) and (2.72) depends on the higher-order modes. Adequate approximations for these contributions are needed to find an impedance Z_c (or a value of δ) which can be described in terms of the propagating mode only.

As soon as one of the equations (2.71) and (2.72) is satisfied, the other one will also be satisfied because the voltage–current pairs on both sides of the equality signs satisfy the circuit equation for the load. This also implies that we cannot introduce independent approximations in both equations. An approximation introduced in one equation will determine the approximation in the other equation.

In working out the equalities (2.71) and (2.72) we will draw a distinction between a TEM mode and a hybrid mode.

2.4.3.1 *TEM mode*   The natural approximation, if we start with the voltage equation (2.71), is that we neglect the contribution of the higher-order modes. In that case (2.71) becomes

$$A\,e^{-j\beta L}(1 + R_{V,L}) = A\,e^{-j\beta L}(1 + R_{V,L})\int_P \mathbf{E}_t \cdot d\mathbf{l}. \tag{2.73}$$

The left- and right-hand side of (2.73) can be equal only if the electric field integral equals 1:

$$\int_P \mathbf{E}_t \cdot d\mathbf{l} = 1. \tag{2.74}$$

Substitution in (2.74) of the gradient form (2.4) of the transversal electric field leads to a new condition:

$$\phi_1 - \phi_0 = \Delta\phi = 1 \tag{2.75}$$

which expresses the fact that the potential difference $\Delta\phi$ introduced in (2.16) is 1 volt. As the electric field of a TEM mode is transversal, the choice of the integration path P does not affect the condition (2.75). This means that the integration path P can be replaced by each path P' which connects the inner and the outer conductor of the waveguide. The combination of (2.75) with (2.37), which follows from the requirement that the circuit model is also an equivalent transmission line, yields the constant χ:

$$\chi = -\frac{1}{\oint_c \frac{\partial \phi}{\partial n} dc}. \tag{2.76}$$

Remember that χ is real as discussed after (2.37), hence $\chi^* = \chi$. Using (2.76), (2.4), and (2.8) we can easily check that the integration of the transversal magnetic field along the contour c is also equal to 1:

$$\oint_c \mathbf{H}_t \cdot \mathbf{dl} = 1. \tag{2.77}$$

The contour c, used in the calculation of the current, can be replaced by any contour c' located between the inner and outer conductor of the waveguide. This implies that in the TEM case c_c in (2.70) and (2.72) can also be replaced by c.

Substitution of (2.77) in (2.72) then shows that the contribution of the higher-order modes to the current flowing through the load is also zero. Once we neglect the higher-order mode terms in either the voltage or the current we must neglect the contribution to the other one. This does not mean that the higher-order modes have no effect. It means that we can neglect their effect on the current and the voltage simultaneously, that is, with the same order of approximation.

The choice (2.76) for χ selects one of the equivalent field transmission lines as circuit transmission line. The circuit voltage V_c and circuit current I_c in the circuit model of the waveguide are given by

$$V_c = V^+ e^{-j\beta x} + V^- e^{j\beta x}$$
$$I_c = \frac{1}{Z_c}(V^+ e^{-j\beta x} - V^- e^{j\beta x}). \tag{2.78}$$

2.4 APPLICATION OF THE CIRCUIT PRINCIPLE

From (2.11) and (2.76) the circuit impedance Z_c is found to be

$$Z_c = -\sqrt{\frac{\mu}{\varepsilon}} \frac{1}{\oint_c \frac{\partial \phi}{\partial n} dc}. \tag{2.79}$$

The coefficients V^+ and V^- represent the amplitudes of the forward and backward propagating circuit voltage waves. The impedance follows from each of the following three equivalent definitions.

(1) the voltage–current definition:

$$Z_c = \frac{V^+}{I^+}; \tag{2.80}$$

(2) the power–voltage definition:

$$Z_c = \frac{|V^+|^2}{2(P^+)^*}; \tag{2.81}$$

(3) the power–current definition:

$$Z_c = \frac{2P^+}{|I^+|^2}. \tag{2.82}$$

For a given TEM mode propagating in the positive direction the three physical quantities, that is, the propagated power P^+, the voltage V^+, and the current I^+, follow from the field $[\mathbf{E}^+, \mathbf{H}^+]$ in the cross-section $x = 0$ of the waveguide associated with the mode propagating in the positive x-direction:

$$V^+ = \int_P \mathbf{E}^+(x=0, y, z) \cdot d\mathbf{l} \tag{2.83}$$

$$I^+ = \oint_c \mathbf{H}^+(x=0, y, z) \cdot d\mathbf{l} \tag{2.84}$$

$$P^+ = \frac{1}{2} \int_S \int [\mathbf{E}^+(x=0, y, z) \times \mathbf{H}^{+*}(x=0, y, z)] \cdot \mathbf{1}_x \, dy \, dz. \tag{2.85}$$

The voltage V^+ represents the voltage difference between the inner and the outer conductor. The current I^+ is the total longitudinal current flowing on the inner conductor. It is important to remark that the fields $\mathbf{E}^+$ and $\mathbf{H}^+$ are total fields as defined in (2.1) but restricted to a mode propagating in the positive x-direction.

The propagation constant β and the characteristic impedance Z_c of the circuit model can be expressed in terms of the capacitance and the inductance of the waveguide. The capacitance C per unit length of the inner conductor with respect to the outer conductor is defined as the charge Q per unit length on the inner conductor divided by the voltage difference V between the two conductors:

$$Q = CV. \tag{2.86}$$

Using the Gauss theorem we find the following value for the capacitance:

$$C = -\varepsilon \oint_c \frac{\partial \phi}{\partial n} dc. \tag{2.87}$$

The inductance L of the waveguide describes the relation between the flux Φ of the magnetic induction per unit length in the x-direction through the path P (or for that matter through any other path P' connecting the inner and the outer conductor) and the current flowing on the inner conductor:

$$\Phi = LI. \tag{2.88}$$

Starting from (2.88), (2.4), (2.8), and (2.76) the following value for L is obtained:

$$L = -\frac{\mu}{\oint_c \frac{\partial \phi}{\partial n} dc}. \tag{2.89}$$

Using the equations (2.79) for the characteristic impedance and (2.22) for the propagation constant and the relations (2.87) and (2.89), the propagation constant and the characteristic impedance can be expressed in terms of L and C:

$$\beta = \omega\sqrt{LC} = \omega\sqrt{\varepsilon\mu} \tag{2.90}$$

and

$$Z_c = \sqrt{\frac{L}{C}} = \sqrt{\frac{\mu}{\varepsilon}}. \tag{2.91}$$

(2.90) and (2.91) are indeed the familiar values to be expected for a coaxial line. Calculation of the capacitance and inductance determines the circuit model of a waveguide propagating a single TEM mode.

2.4.3.2 *Hybrid mode* For a TEM mode we showed that the choice of the continuity equations (2.71) or (2.72) from which we start our approximations, neglecting the contribution of the higher-order modes, does not affect the circuit model and its characteristic impedance. The characteristic impedance can be calculated from the propagated power P^+, the voltage V^+, and the current I^+ associated with the TEM mode in three equivalent ways. This

2.4 APPLICATION OF THE CIRCUIT PRINCIPLE

will not be the case for a hybrid waveguide as we cannot neglect the higher-order modes in both continuity equations simultaneously.

If we start from the voltage continuity equation (2.71) and neglect the higher-order mode contribution, the circuit voltage will equal the line voltage associated with the fundamental hybrid mode. But the circuit current will differ from the line current of the fundamental hybrid mode. The impedance of the resulting circuit model can be calculated from the power P^+ and the voltage V^+ of the fundamental hybrid mode. Therefore we call it a power–voltage defined impedance. The circuit model has three equivalent impedance definitions but the two others, that is, the power–current and the voltage–current, are derived from the first. Unlike the circuit model of a TEM mode, the power–voltage based circuit model will also depend on the choice of the path P.

Similarly, if we start from the current continuity equation (2.72), the impedance of the circuit model follows from the power P^+ and the current I^+ and is called a power–current defined impedance. The two other equivalent impedance definitions associated with the circuit model are derived from the first one. The circuit model depends on the choice of the contour c selected for the current defnition. Only the contour c_c which describes the circumference seems to have a clear-cut physical meaning because all the other contours result in a total current which consists of two contributions: the total longitudinal current flowing on the conductor and the dielectric displacement current through the surface enclosed by the contour c.

If we drop the requirement that the circuit model is an equivalent transmission line model we can neglect the effect of the higher-order modes entirely. The circuit model is built up with the voltage V^+ and the current I^+ of the fundamental mode and the impedance is a voltage–current defined impedance. But the circuit does not propagate the same power as the waveguide. The usefulness of any voltage–current defined impedance is limited but we will discuss the model here because it was often used in the design of high-frequency components. We will discuss this in more detail in Section 2.4.3.2.4.

We will now investigate the three different circuit models for hybrid waveguides in more detail. Subsequently, we will discuss the best choice of a circuit model for the most common high-frequency transmission lines, such as the microstrip, and give a short overview of the different models that have been published in the literature. Finally, we will illustrate the difference between the three impedance definitions for the single microstrip line.

2.4.3.2.1 *Power–voltage definition* If we start from the same approximation as in the TEM case, that is, if we neglect the effect of the higher-order modes on the line voltage over the load in (2.71), we find exactly the same condition as in the TEM case for the transversal electric field of the fundamental mode:

$$\int_P \mathbf{E}_t \cdot d\mathbf{l} = 1. \qquad (2.92)$$

But now the integral depends on the choice of path P because (in general) the hybrid mode has a longitudinal magnetic field.

As discussed at the beginning of Section 2.4.3.1 the transversal electric field $\mathbf{E}_t(y, z)$ associated with an equivalent transmission line can always be multiplied by an arbitrary complex constant δ. The condition (2.92) determines the value of δ, of all the field components, and of the impedance of the circuit model. The impedance can be calculated from (2.41), using the transversal electric field $\mathbf{E}_t$ satisfying (2.92):

$$Z_c = \frac{-\omega\beta}{\int\int_S \left[\frac{1}{\mu}(\nabla_t \times \mathbf{E}_t) \cdot (\nabla_t \times \mathbf{E}_t)^* - \omega^2 \varepsilon |\mathbf{E}_t|^2\right] dy\, dz}. \qquad (2.93)$$

Taking into account condition (2.94) the circuit voltage V_c is given by

$$V_c = V^+ e^{-j\beta x} + V^- e^{j\beta x} \qquad (2.94)$$

where V^+ and V^- represent the values of the line voltage along path P in the cross-section $x = 0$ of the electric field propagating in the forward and backward direction respectively:

$$V^\pm = \int_P \mathbf{E}^\pm(x = 0, y, z) \cdot d\mathbf{l}. \qquad (2.95)$$

The circuit current I_c follows from the circuit voltage (2.94) and the circuit or characteristic impedance (2.93):

$$I_c = \frac{1}{Z_c}(V^+ e^{-j\beta x} - V^- e^{j\beta x}). \qquad (2.96)$$

Equations (2.94) and (2.96) describe a circuit model of a hybrid waveguide.

We will now show that, unlike the circuit current in the TEM case, the circuit current I_c cannot be expressed in terms of the line currents I^+ and I^- in the cross-section $x = 0$ of the forward and backward propagating fundamental mode. These currents I^+ and I^- follow from the integration along the contour c_c of the magnetic field at $x = 0$:

$$I^\pm = \oint_{c_c} \mathbf{H}^\pm(x = 0, y, z) \cdot d\mathbf{l}. \qquad (2.97)$$

For a TEM mode, the condition (2.92) also implied that the higher-order

2.4 APPLICATION OF THE CIRCUIT PRINCIPLE

mode contributions to the current through the load could be neglected together with the higher-order mode contribution to the voltage through the load. This is a consequence of the fact that the line integral (2.77) of the magnetic field along the contour c equals 1, independent of the choice of path P and contour c. For a hybrid mode the line integral (2.77) depends on the choice of the contour c, because of the presence of a longitudinal electric field, and on the choice of the path P, because this path determines the amplitude of the transversal electric field through the condition (2.92). In general the line integral (2.77) will differ from 1 and now the higher-order modes will contribute to the total line current through the load in the circuit model. Using (2.70) and (2.97) we can determine this condition as the difference between the total line current through the load and the contribution of the fundamental mode:

$$I_L - (I^+ e^{-j\beta L} - I^- e^{j\beta L}) = \sum_{i=1}^{N} -\frac{R_i A}{Z_i} e^{j\beta_i L} \oint_{c_c} \mathbf{H}_{t,i} \cdot d\mathbf{l}. \quad (2.98)$$

If we denote the amplitude of the forward and backward propagating current waves in the circuit by $I_c^+ = V^+/Z_c$ and $I_c^- = V^-/Z_c$ respectively (see also (2.96)) then we have proved that

$$I_c^\pm \neq I^\pm. \quad (2.99)$$

The circuit current is not the line current associated with the fundamental mode, but is such that at the load interface it represents the total line current flowing in the load. The accuracy of this representation depends on the approximation (2.92) made in the voltage calculation.

As the circuit model is also an equivalent transmission line it propagates the same power as the fundamental mode. From the expressions (2.94) and (2.96) we can find the power P^+ propagated by the forward propagating mode:

$$P^+ = \frac{1}{2} \frac{|V^+|^2}{Z_c^*}. \quad (2.100)$$

Both the power P^+ and the voltage V^+ follow from the solution of the mode problem, so we can use (2.100) instead of (2.93) to calculate the characteristic impedance Z_c:

$$Z_c = \frac{|V^+|^2}{2(P^+)^*}. \quad (2.101)$$

This is the power–voltage definition of the characteristic impedance. The other equivalent definitions of Z_c, the power–current and the voltage–current, are based upon the circuit current I_c^+ and are respectively given by:

$$Z_c = \frac{V^+}{I_c^+} \tag{2.102}$$

and

$$Z_c = \frac{2P^+}{|I_c^+|^2}. \tag{2.103}$$

Only the first definition can be derived from the fundamental mode only. Therefore we call the impedance in this circuit model a power–voltage defined impedance. But within the circuit model, there exist three equivalent definitions (2.101)–(2.103) for the impedance.

2.4.3.2.2 *Power–current definition* In the previous section we neglected the contribution of higher-order modes in the voltage continuity equation (2.71) which led to the power–voltage defined impedance. In this section we will start from the other continuity equation, that is, the current continuity equation (2.72). Neglecting the contribution of the higher-order modes to the total current through the load leads to:

$$\oint_{c_c} \mathbf{H}_t \cdot \mathbf{dl} = 1. \tag{2.104}$$

This condition, in combination with the equivalent transmission line condition (2.42), determines the characteristic impedance of, and all the field components associated with, the circuit model. The circuit current I_c is given by

$$I_c = I^+ e^{-j\beta x} - I^- e^{j\beta x}. \tag{2.105}$$

The circuit voltage follows from the circuit current and the characteristic impedance:

$$V_c = Z_c(I^+ e^{-j\beta x} + I^- e^{j\beta x}). \tag{2.106}$$

The circuit model propagates the correct power so the impedance Z_c can be calculated from the propagated power P^+ and the current I^+ of a forward travelling mode:

$$Z_c = \frac{2P^+}{|I^+|^2}. \tag{2.107}$$

From reasoning similar to that for the power–voltage defined impedance in Section 2.4.3.2.1 we can conclude that the circuit voltage and the line voltage built up by the fundamental mode at the load are not equal and that

2.4 APPLICATION OF THE CIRCUIT PRINCIPLE

the difference describes the higher-order modes contribution to the voltage. From (2.69) and (2.95) we find

$$V_L - (V^+ e^{-j\beta L} + V^- e^{j\beta L}) = \sum_{i=1}^{N} R_i A\, e^{j\beta_i L} \int_P \mathbf{E}_{t,i} \cdot d\mathbf{l}. \quad (2.108)$$

The value of this voltage difference is only an approximation because we neglected the effect of the higher-order modes on the current. This results in a modified contribution of the higher-order modes to the voltage such that the circuit model remains an equivalent transmission line and satisfies all the boundary conditions.

Besides the power–current definition, the characteristic impedance Z_c has two other equivalent definitions, that is, the power–voltage and the voltage–current definitions:

$$Z_c = \frac{V_c^+}{I^+} \quad (2.109)$$

and

$$Z_c = \frac{|V_c^+|^2}{2(P^+)^*}. \quad (2.110)$$

Both definitions are based upon the voltage $V_c^+ = Z_c I^+$ (see (2.106)) which does not follow directly from the analysis of the fundamental hybrid mode; it is a derived quantity. Therefore we say that the impedance of this circuit model is based upon a power–current definition.

2.4.3.2.3 Voltage–current definition The difference between the power–voltage and the power–current defined characteristic impedance is a consequence of the fact that the effect of the higher-order modes on the circuit voltage and current cannot be neglected simultaneously. Only two conditions, one of which is the equivalent transmission line condition (2.41) or (2.42), can be satisfied together. However, if we drop the condition (2.41) we can impose both conditions (2.92) and (2.104) arising from neglecting the higher-order modes in the circuit voltage and current respectively. The circuit equations are now

$$V_c = V^+ e^{-j\beta x} + V^- e^{j\beta x} \quad (2.111)$$

and

$$I_c = I^+ e^{-j\beta x} - I^- e^{j\beta x}. \quad (2.112)$$

The characteristic impedance follows from the voltage V^+ and the current I^+:

$$Z_c = \frac{V^+}{I^+}. \quad (2.113)$$

The power $P_c = V^+(I^+)^*/2$ propagated by the circuit differs from the power P propagated in the waveguide because the circuit model is no longer an equivalent transmission line model. It is obvious that the usefulness of this voltage–current model is restricted. The two other equivalent impedance definitions are based upon the power P_c^+ propagated by the forward travelling wave in the circuit:

$$Z_c = \frac{|V^+|^2}{2(P_c^+)^*} \qquad (2.114)$$

and

$$Z_c = \frac{2P_c^+}{|I^+|^2}. \qquad (2.115)$$

2.4.3.2.4 Discussion The impedance of a waveguide is used to make circuit interconnections between a waveguide and TEM elements while the waveguide is not necessarily a TEM structure itself. The impedance describes the power exchange between the waveguide and its load and driver. The ability to predict the performance of a circuit will depend partially on the accuracy of the impedance definition of the waveguides in the circuit.

For TEM waveguides we found a unique impedance. A TEM waveguide propagates a fundamental TEM mode. The voltage and the current in the circuit model can be interpreted respectively as the line voltage and line current of the fundamental mode.

For a hybrid mode we introduced three different definitions of the characteristic impedance. The first two, the power–voltage and the power–current, are based upon the power equivalence principle. The third one, the voltage–current, does not require that the circuit model propagates the same power as the fundamental mode. The circuit current and voltage can no longer be expressed in terms of the usual TEM path integrals of the fields.

The key question is which choice of the impedance of a hybrid waveguide is the best TEM equivalent. The choice of the impedance depends on the type of waveguide and the load and driver of the waveguide under study. In Fig. 2.7 we sketch some waveguides frequently used in high-frequency/high-speed circuit design: (a) the microstrip; (b) the slotline; (c) the coplanar waveguide; (d) coplanar strips; and (e) the stripline.

We will now discuss the choice of the impedance for some of the waveguides depicted in Fig. 2.7.

Several definitions for the characteristic impedance of the microstrip have been introduced in the literature. The choice of the impedance of the microstrip has always been a subject of discussion before Brews introduced a consistent model which is based upon the power equivalence and circuit principles discussed in Chapter 1 ([1] and [3]). We will now give a brief chronological overview of some publications which discussed the characteristic impedance of the microstrip.

2.4 APPLICATION OF THE CIRCUIT PRINCIPLE

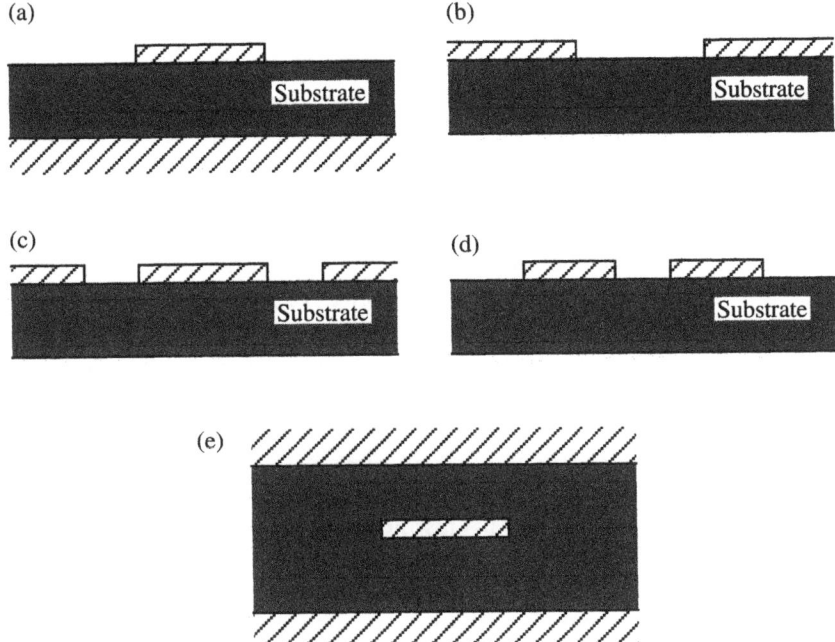

Fig. 2.7. Frequently used waveguides in high-frequency/high-speed circuit design: (a) the microstrip; (b) the slotline; (c) the coplanar waveguide; (d) coplanar strips; (e) the stripline.

Bianco et al. [4] specified five definitions of the impedance based on three different definitions of the voltage and the current. The impedances have the same value for the d.c. limit but have a strongly different frequency behaviour. Bianco used Getsinger's model [5] which approximates the fundamental mode by a *longitudinal section electric* (LSE) mode.

Getsinger [5] applied the concept of the wave impedance to the microstrip. Within the LSE mode approximation he finds a unique and unambiguous impedance because the wave impedance, related to the impedance and first introduced by Schelkunoff [2], is the same at every point of the cross-section of the microstrip. He found that only one definition of Bianco et al. [4] is consistent with the wave impedance based approach. The impedance is proportional to the wave impedance and the proportionality factor is such that the impedance has the correct quasi-static value. In the LSE model the electric field is transversal while the magnetic field has a longitudinal component. Therefore the voltage is path dependent while the current is path independent. Thus a unique current can be defined in the LSE model, but not a unique voltage. Getsinger [5] introduced an 'apparent' LSE power–current defined impedance which describes the power exchange between the microstrip and a TEM line.

Jansen and Kirschning [6] give several arguments which favour the power–current defined impedance over the other definitions. Firstly, a correct description of the wave amplitudes at a microstrip junction or any microstrip excitation requires that the power be part of the impedance definition. Both the power–voltage and the power–current definitions could be used. Secondly, the longitudinal current is almost constant as a function of frequency for low frequencies and constant values of the propagated power. This can be explained by the presence of the substrate–air interface which has a first-order effect on the electric field and only a second-order effect on the magnetic field and the current for increasing frequency. The fact that the power–current defined impedance remains almost constant for low frequencies makes this definition a very suitable one to serve as a TEM equivalent of a microstrip.

The slotline is a typical example of a waveguide which does not propagate a fundamental mode for low frequencies, that is, the fundamental mode has a cut-off frequency different from 0 Hz. The mode of propagation is almost transversal electric in nature [7]. This means that the line voltage is almost path independent. Therefore a power–voltage defined impedance is preferable over a power–current defined impedance.

The stripline is a waveguide with only one dielectric medium; the structure is homogeneous and therefore it propagates a TEM mode. The three definitions of the impedance are identical.

2.4.3.2.5 Comparison between the three impedance definitions for a microstrip
In this section we will show the difference between the three definitions of the characteristic impedance for a microstrip line. The configuration under study is depicted in Fig. 2.8. The substrate has a relative permittivity $\varepsilon_r = 11.7$ and a thickness $h = 3.17$ mm. The strip and the ground plane are infinitely thin and perfectly conducting. The strip has a width $w = 3.05$ mm. Due to its layered character the microstrip propagates a hybrid mode. We analysed the structure in the frequency range 0–15 GHz. The effective relative permittivity $\varepsilon_{r,\text{eff}} = (\beta/k_0)^2$ and the three characteristic impedances are shown in Fig. 2.9 as a function of frequency.

The value of the effective relative permittivity for a certain frequency can be read off the left-hand vertical axis and the values of the characteristic

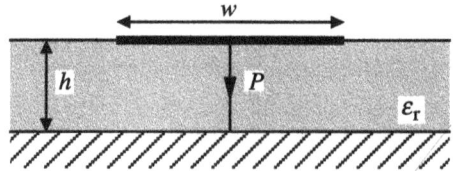

Fig. 2.8. A microstrip transmission line.

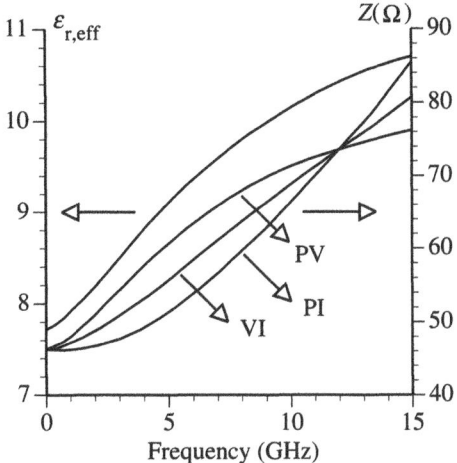

Fig. 2.9. Effective relative permittivity and the three characteristic impedances as a function of frequency for the configuration of Fig. 2.8.

impedances can be read off the right-hand vertical axis. For a detailed discussion of the behaviour of the effective relative permittivity we refer the reader to Chapter 7. Here we are only interested in the behaviour of the characteristic impedance. The three definitions are based upon the propagated power P, the total longitudinal conduction current I, and the centre strip voltage V. The path P used in the calculation is located in the centre of the strip and is depicted in Fig. 2.8. From Fig. 2.9 we can see that the three definitions coincide in the low-frequency limit. This confirms that for low frequencies the quasi-static approximation of the total field is valid. As the frequency increases, the impedance value increases for all three definitions but the behaviour is different.

The figure also shows that the power–current (PI) defined characteristic impedance has the most TEM-like character as its value only starts to increase at higher frequencies as compared to the power–voltage (PV) and voltage–current (VI) defined impedances.

For a particular frequency (around 12 GHz) the three definitions result in the same impedance value. For this frequency the voltage and the current associated with the fundamental mode describe the propagated power in the microstrip correctly. For another choice of the path P or the contour c we would have found a different intersection frequency.

References

[1] Brews, J. R. (1986). Transmission line models for lossy waveguide interconnections in VLSI. *IEEE Transactions on Electronic Devices*, **ED-33**, 9, 1356–1365.

[2] Schelkunoff, S. A. (1943). *Electromagnetic Waves.* Van Nostrand: New York.
[3] Brews, J. R. (1987). Characteristic impedance of microstrip lines. *IEEE Transactions on Microwave Theory and Techniques*, **MTT-35**, 1, 30–34.
[4] Bianco, B., Panini, L., Parodi, M., and Ridella, S. (1978). Some considerations about the frequency dependence of the characteristic impedance of uniform microstrips. *IEEE Transactions on Microwave Theory and Techniques*, **MTT-26**, 3, 182–186.
[5] Getsinger, W. J. (1983). Measurement and modeling of the apparent characteristic impedance of microstrip. *IEEE Transactions on Microwave Theory and Techniques*, **MTT-31**, 8, 624–632.
[6] Jansen, R. H. and Kirschning, M. (1983). Arguments and an accurate model for the power-current formulation of microstrip characteristic impedance. *Arch. Elek. Übertragung*, **37**, 3/4, 108–112.
[7] Cohn, S. B. (1969). Slot line on a dielectric substrate. *IEEE Transactions on Microwave Theory and Techniques*, **MTT-17**, 10, 768–778.

3
CIRCUIT DESCRIPTION OF TWO COUPLED LOSSLESS WAVEGUIDE STRUCTURES

3.1 Introduction

In Chapter 2 the circuit model for a general waveguide structure supporting a single fundamental mode was derived. The circuit model is based upon the power equivalence and circuit principles. In the present chapter these two principles will be applied to construct the circuit model of two coupled lossless waveguide structures. The extension of the approach for single waveguide structures to two coupled waveguide structures was first presented in [1]. The general case of an arbitrary number of coupled lossy or lossless waveguide structures will be discussed in Chapter 4.

The different steps leading to a circuit model are the same as for the single waveguide structure. The theory will be exemplified by means of a coupled microstrip structure which is described in Section 3.2. The starting point is the electromagnetic field representation of the two fundamental modes propagated by the structure under study. In Section 3.3 a general coupled field transmission line representation for the coupled waveguide structure is derived along with the field components associated with that transmission line. In Section 3.4 the subset of coupled field transmission lines which are also equivalent coupled transmission lines are found by applying the power equivalence principle. The final step, the application of the circuit principle, is elaborated in Section 3.5 for a power–current defined impedance. From the current definition follow the line-mode impedances and the circuit voltages. In Section 3.5 the power distribution over the conductors of the waveguide is determined and the difference between this and the classical definition of the power distribution is discussed. We conclude this chapter with an example illustrating the difference between the classical power distribution definition and the power distribution found in the self-consistent circuit model presented in this chapter.

3.2 Two coupled lossless hybrid waveguide structures

Figure 3.1 shows the cross-section of two coupled microstrips. This is a typical hybrid waveguide which propagates two fundamental modes. For low

3 TWO COUPLED LOSSLESS WAVEGUIDE STRUCTURES

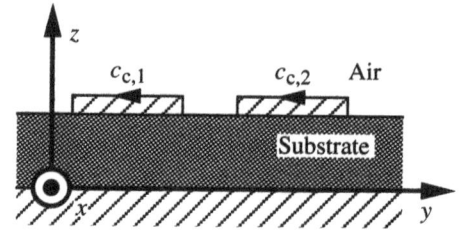

Fig. 3.1. Cross-section of two coupled microstrip lines.

frequencies, only those two modes are propagated by the structure. All the other modes have a cut-off frequency.

The two strips are placed on top of the interface between the substrate and the air layer. The double-layered medium is backed by a ground plane. As we are restricting the analysis to lossless structures the layers are characterized by a real permittivity and permeability and the strips and ground plane are perfectly conducting. A strip, or in general a conductor, is denoted by the subscript p, where p takes the values 1 or 2. A mode and the associated quantities are denoted by a superscript (i) $(i = 1, 2)$.

3.3 Modal field representation

The total field $[\mathbf{E}, \mathbf{H}]$ propagating in the waveguide structure can be written as the sum of the fields of the two fundamental modes:

$$\mathbf{E} = \mathbf{E}_{tot}^{(1)} + \mathbf{E}_{tot}^{(2)}$$
$$\mathbf{H} = \mathbf{H}_{tot}^{(1)} + \mathbf{H}_{tot}^{(2)}. \tag{3.1}$$

The general form of a modal field $[\mathbf{E}_{tot}^{(i)}, \mathbf{H}_{tot}^{(i)}]$ $(i = 1, 2)$ has already been derived for a waveguide propagating a single fundamental mode. The form (2.28) in the previous chapter remains valid for coupled structures. Substitution in (3.1) gives

$$\mathbf{E} = V_f^{(1)}(x)\mathbf{E}_t^{(1)}(y, z) + R_0 I_f^{(1)}(x)\mathbf{E}_l^{(1)}(y, z)$$
$$+ V_f^{(2)}(x)\mathbf{E}_t^{(2)}(y, z) + R_0 I_f^{(2)}(x)\mathbf{E}_l^{(2)}(y, z) \tag{3.2}$$

for the electric field and

$$\mathbf{H} = I_f^{(1)}(x)\mathbf{H}_t^{(1)}(y, z) + \frac{V_f^{(1)}(x)}{R_0} \mathbf{H}_l^{(1)}(y, z)$$
$$+ I_f^{(2)}(x)\mathbf{H}_t^{(2)}(y, z) + \frac{V_f^{(2)}(x)}{R_0} \mathbf{H}_l^{(2)}(y, z) \tag{3.3}$$

for the magnetic field.

3.3 MODAL FIELD REPRESENTATION

The fields $\mathbf{E}_t^{(i)}$, $\mathbf{H}_t^{(i)}$, $\mathbf{E}_l^{(i)}$, and $\mathbf{H}_l^{(i)}$ of each mode i follow from the eigenvalue equations (2.31) and (2.32) presented in the previous chapter for the propagation constant $\beta^{(i)}$ and from the associated curl and divergence equations (2.29) and (2.30) of Maxwell. As we are restricting our attention to lossless waveguides the transversal and longitudinal fields have a phase which is in quadrature at each point of the cross-section. Moreover, each field $\mathbf{E}_t^{(i)}$, $\mathbf{H}_t^{(i)}$, $\mathbf{E}_l^{(i)}$, and $\mathbf{H}_l^{(i)}$ has a fixed phase in a cross-section of the waveguide. This phase property will be used in the subsequent discussion.

The field voltage $V_f^{(i)}$ and the field current $I_f^{(i)}$ of mode i are a solution to the classical transmission line equations for the voltage and the current on a transmission line:

$$\frac{dV_f^{(i)}(x)}{dx} + j\beta^{(i)}Z_f^{(i)}I_f^{(i)}(x) = 0$$

$$\frac{dI_f^{(i)}(x)}{dx} + \frac{j\beta^{(i)}}{Z_f^{(i)}} V_f^{(i)}(x) = 0. \tag{3.4}$$

The field transmission line of mode i governed by the equations (3.4) is characterized by a propagation constant $\beta^{(i)}$ and a field impedance $Z_f^{(i)}$. The propagation constant $\beta^{(i)}$ is the propagation constant of the mode. The impedance $Z_f^{(i)}$ is still an arbitrary complex number. A general solution consists of two waves propagating in opposite directions along the x-axis:

$$V_f^{(i)}(x) = A^{(i)} e^{-j\beta^{(i)}x} + B^{(i)} e^{j\beta^{(i)}x}$$

$$I_f^{(i)}(x) = \frac{1}{Z_f^{(i)}} (A^{(i)} e^{-j\beta^{(i)}x} - B^{(i)} e^{j\beta^{(i)}x}). \tag{3.5}$$

The amplitudes $A^{(i)}$ and $B^{(i)}$ follow from the boundary conditions at the beginning and the end of the waveguide.

The two equations in (3.4) describe the field voltage and current associated with a mode i and not with a conductor p. However, for the derivation of a circuit model of the two fundamental modes we need a coupled field transmission line model for the two conductors. In each such coupled transmission line model the voltage and the current of a conductor will respectively describe the longitudinal variation of the transversal electric and magnetic field associated with that conductor. Therefore, the field voltage (current) of a conductor can always be written as a linear combination of the field voltages (currents) of the modes. The field voltage and current of conductor p are denoted by $V_{f,p}(x)$ and $I_{f,p}(x)$ respectively, $p = 1, 2$. For the field voltages this linear relationship is written in the following form:

$$V_{f,p}(x) = v_p^{(1)}V_f^{(1)}(x) + v_p^{(2)}V_f^{(2)}(x). \tag{3.6}$$

3 TWO COUPLED LOSSLESS WAVEGUIDE STRUCTURES

A coefficient $v_p^{(i)}$ describes the contribution of the field voltage $V_f^{(i)}$ of mode i to the field voltage $V_{f,p}$ of conductor p. The coefficients $v_p^{(i)}$ are still unknown. A similar relationship holds for the field currents of the conductors and the field currents of the modes:

$$I_{f,p}(x) = i_p^{(1)} I_f^{(1)}(x) + i_p^{(2)} I_f^{(2)}(x). \tag{3.7}$$

The coefficients $i_p^{(i)}$ are also unknown.

The electromagnetic field excited by the two fundamental modes can also be expressed in terms of the field voltages $V_{f,p}(x)$ and currents $I_{f,p}(x)$ associated with the conductors. For the electric field, we invert the equations (3.6) and (3.7) and replace the field voltages $V_f^{(i)}$ and $I_f^{(i)}$ of the modes with the field voltages $V_{f,p}(x)$ and currents $I_{f,p}(x)$ of the conductors. After reordering, we find the following expressions for the total electric field propagating in the waveguide structure:

$$\mathbf{E} = \mathbf{E}_{t,1} V_{f,1} + \mathbf{E}_{t,2} V_{f,2} + R_0 \mathbf{E}_{l,1} I_{f,1} + R_0 \mathbf{E}_{l,2} I_{f,2} \tag{3.8}$$

with the electric field components in (3.8) given by

$$\begin{aligned}
\mathbf{E}_{t,1}(y,z) &= \frac{v_2^{(2)} \mathbf{E}_t^{(1)}(y,z) - v_2^{(1)} \mathbf{E}_t^{(2)}(y,z)}{\Delta v} \\
\mathbf{E}_{t,2}(y,z) &= \frac{-v_1^{(2)} \mathbf{E}_t^{(1)}(y,z) + v_1^{(1)} \mathbf{E}_t^{(2)}(y,z)}{\Delta v} \\
\mathbf{E}_{l,1}(y,z) &= \frac{i_2^{(2)} \mathbf{E}_l^{(1)}(y,z) - i_2^{(1)} \mathbf{E}_l^{(2)}(y,z)}{\Delta i} \\
\mathbf{E}_{l,2}(y,z) &= \frac{-i_1^{(2)} \mathbf{E}_l^{(1)}(y,z) + i_1^{(1)} \mathbf{E}_l^{(2)}(y,z)}{\Delta i}
\end{aligned} \tag{3.9}$$

with

$$\Delta i = i_1^{(1)} i_2^{(2)} - i_1^{(2)} i_2^{(1)} \tag{3.10}$$

and

$$\Delta v = v_1^{(1)} v_2^{(2)} - v_1^{(2)} v_2^{(1)}. \tag{3.11}$$

An analogous expression can be found for the magnetic field:

$$\mathbf{H} = \mathbf{H}_{t,1} I_{f,1} + \mathbf{H}_{t,2} I_{f,2} + \frac{\mathbf{H}_{l,1}}{R_0} V_{f,1} + \frac{\mathbf{H}_{l,2}}{R_0} V_{f,2} \tag{3.12}$$

3.3 MODAL FIELD REPRESENTATION

with the magnetic field components in (3.12) given by

$$\mathbf{H}_{t,1}(y,z) = \frac{i_2^{(2)}\mathbf{H}_t^{(1)}(y,z) - i_2^{(1)}\mathbf{H}_t^{(2)}(y,z)}{\Delta i}$$

$$\mathbf{H}_{t,2}(y,z) = \frac{-i_1^{(2)}\mathbf{H}_t^{(1)}(y,z) + i_1^{(1)}\mathbf{H}_t^{(2)}(y,z)}{\Delta i}$$

$$\mathbf{H}_{l,1}(y,z) = \frac{v_2^{(2)}\mathbf{H}_l^{(1)}(y,z) - v_2^{(1)}\mathbf{H}_l^{(2)}(y,z)}{\Delta v}$$

$$\mathbf{H}_{l,2}(y,z) = \frac{-v_1^{(2)}\mathbf{H}_l^{(1)}(y,z) + v_1^{(1)}\mathbf{H}_l^{(2)}(y,z)}{\Delta v}.$$

(3.13)

The two sets of differential equations governing the longitudinal dependence of the field voltage $V_{f,p}(x)$ and the current $I_{f,p}(x)$ of the conductors are found in the following way. For the first set, we start from (3.6) and first take the derivative with respect to x of both sides and then replace the derivatives of the field voltages $V_f^{(i)}$ of the modes in the right-hand side by the field currents $I_f^{(i)}$, using (3.4). In the resulting equation we replace the field currents $I_f^{(i)}$ of the modes by the field currents $I_{f,p}(x)$ of the conductors by inverting the equations (3.7). The result of these simple steps is a first set of equations relating the first derivatives of the field voltages with the field currents of the conductors:

$$\frac{dV_{f,p}(x)}{dx} = \frac{-j\beta^{(1)}Z_f^{(1)}v_p^{(1)}i_2^{(2)} + j\beta^{(2)}Z_f^{(2)}v_p^{(2)}i_2^{(1)}}{\Delta i} I_{f,1}$$

$$+ \frac{j\beta^{(1)}Z_f^{(1)}v_p^{(1)}i_1^{(2)} - j\beta^{(2)}Z_f^{(2)}v_p^{(2)}i_1^{(1)}}{\Delta i} I_{f,2}.$$ (3.14)

If we repeat the same steps but start from (3.7) we find the second, dual, set of equations:

$$\frac{dI_{f,p}(x)}{dx} = \frac{-j\dfrac{\beta^{(1)}}{Z_f^{(1)}}i_p^{(1)}v_2^{(2)} + j\dfrac{\beta^{(2)}}{Z_f^{(2)}}i_p^{(2)}v_2^{(1)}}{\Delta v} V_{f,1}$$

$$+ \frac{j\dfrac{\beta^{(1)}}{Z_f^{(1)}}i_p^{(1)}v_1^{(2)} - j\dfrac{\beta^{(2)}}{Z_f^{(2)}}i_p^{(2)}v_1^{(1)}}{\Delta v} V_{f,2}.$$ (3.15)

The equations (3.14) and (3.15) have the same form as the classical equations

3 TWO COUPLED LOSSLESS WAVEGUIDE STRUCTURES

for the voltages and the currents in two coupled transmission lines:

$$\frac{dV_1(x)}{dx} = -z_{11}I_1(x) - z_{12}I_2(x)$$

$$\frac{dV_2(x)}{dx} = -z_{21}I_1(x) - z_{22}I_2(x)$$
(3.16)

and

$$\frac{dI_1(x)}{dx} = -y_{11}V_1(x) - y_{12}V_2(x)$$

$$\frac{dI_2(x)}{dx} = -y_{21}V_1(x) - y_{22}V_2(x).$$
(3.17)

The coefficients z_{pp} and y_{pp} ($p = 1, 2$) are respectively the self-impedance and self-admittance per unit length of line (or conductor) p. The coefficients z_{pq} and y_{pq} ($p, q = 1, 2$ and $p \neq q$) represent respectively the mutual impedance and mutual admittance per unit length.

A general solution to the coupled field transmission line equations (3.14) and (3.15) can be found by substituting the general solution (3.5) of the field transmission line equations of a mode into the equations (3.6) and (3.7):

$$V_{f,p}(x) = v_p^{(1)}(A^{(1)} e^{-j\beta^{(1)}x} + B^{(1)} e^{j\beta^{(1)}x}) + v_p^{(2)}(A^{(2)} e^{-j\beta^{(2)}x} + B^{(2)} e^{j\beta^{(2)}x}) \quad (3.18)$$

and

$$I_{f,p}(x) = \frac{i_p^{(1)}}{Z_f^{(1)}} (A^{(1)} e^{-j\beta^{(1)}x} - B^{(1)} e^{j\beta^{(1)}x}) + \frac{i_p^{(2)}}{Z_f^{(2)}} (A^{(2)} e^{-j\beta^{(2)}x} - B^{(2)} e^{j\beta^{(2)}x}).$$
(3.19)

Both modes contribute to the field voltage $V_{f,p}(x)$ and the field current $I_{f,p}(x)$ on each conductor p.

The voltage coefficients $v_p^{(i)}$, the current coefficients $i_p^{(i)}$, and the field impedances $Z_f^{(i)}$ can be chosen freely. Each choice results in a different coupled field transmission line for the waveguide structure.

3.4 Coupled equivalent transmission lines

By analogy with the single waveguide structure, we expect to find coupled field transmission lines which propagate the same power as the coupled waveguide structure. These lines are called coupled equivalent transmission lines of the structure. They will follow from the equality between the power propagated in the waveguide and the power propagated in the coupled field transmission lines.

In order to find the power propagated by the coupled waveguide structure we calculate the Poynting vector using the expressions (3.8), (3.9), (3.12), and

3.4 COUPLED EQUIVALENT TRANSMISSION LINES

(3.13) for the electric and magnetic fields. Then we integrate the longitudinal component of this vector over the cross-section of the structure. We find:

$$P(x) = \tfrac{1}{2} V_{f,1}(x) I^*_{f,1}(x) \times$$

$$\frac{v_2^{(2)}(i_2^{(2)})^* p^{(1)(1)} - v_2^{(2)}(i_2^{(1)})^* p^{(1)(2)} - v_2^{(1)}(i_2^{(2)})^* p^{(2)(1)} + v_2^{(1)}(i_2^{(1)})^* p^{(2)(2)}}{\Delta v \, \Delta i^*}$$

$$+ \tfrac{1}{2} V_{f,2}(x) I^*_{f,2}(x) \times$$

$$\frac{v_1^{(2)}(i_1^{(2)})^* p^{(1)(1)} - v_1^{(2)}(i_1^{(1)})^* p^{(1)(2)} - v_1^{(1)}(i_1^{(2)})^* p^{(2)(1)} + v_1^{(1)}(i_1^{(1)})^* p^{(2)(2)}}{\Delta v \, \Delta i^*}$$

$$+ \tfrac{1}{2} V_{f,1}(x) I^*_{f,2}(x) \times$$

$$\frac{-v_2^{(2)}(i_1^{(2)})^* p^{(1)(1)} + v_2^{(2)}(i_1^{(1)})^* p^{(1)(2)} + v_2^{(1)}(i_1^{(2)})^* p^{(2)(1)} - v_2^{(1)}(i_1^{(1)})^* p^{(2)(2)}}{\Delta v \, \Delta i^*}$$

$$+ \tfrac{1}{2} V_{f,2}(x) I^*_{f,1}(x) \times$$

$$\frac{-v_1^{(2)}(i_2^{(2)})^* p^{(1)(1)} + v_1^{(2)}(i_2^{(1)})^* p^{(1)(2)} + v_1^{(1)}(i_2^{(2)})^* p^{(2)(1)} - v_1^{(1)}(i_2^{(1)})^* p^{(2)(2)}}{\Delta v \, \Delta i^*}$$

(3.20)

with

$$p^{(i)(j)} = \int\!\!\int_S [\mathbf{E}_t^{(i)} \times (\mathbf{H}_t^{(j)})^*] \cdot \mathbf{1}_x \, dS. \qquad (3.21)$$

For a lossless coupled waveguide structure it can be shown [2], starting from the Lorentz reciprocity theorem, that the modes 1 and 2 are power orthogonal, that is,

$$\int\!\!\int_S [\mathbf{E}_t^{(1)} \times (\mathbf{H}_t^{(2)})^*] \cdot \mathbf{1}_x \, dS = \int\!\!\int_S [\mathbf{E}_t^{(2)} \times (\mathbf{H}_t^{(1)})^*] \cdot \mathbf{1}_x \, dS = 0. \quad (3.22)$$

Hence, the cross power terms $p^{(1)(2)}$ and $p^{(2)(1)}$ in (3.20) vanish. The total power propagated by the structure is the sum of the powers propagated by the individual modes:

$$P(x) = \tfrac{1}{2} V_{f,1}(x) I^*_{f,1}(x) \frac{v_2^{(2)}(i_2^{(2)})^* p^{(1)} + v_2^{(1)}(i_2^{(1)})^* p^{(2)}}{\Delta v \, \Delta i^*}$$

$$+ \tfrac{1}{2} V_{f,2}(x) I^*_{f,2}(x) \frac{v_1^{(2)}(i_1^{(2)})^* p^{(1)} + v_1^{(1)}(i_1^{(1)})^* p^{(2)}}{\Delta v \, \Delta i^*}$$

$$- \tfrac{1}{2} V_{f,1}(x) I^*_{f,2}(x) \frac{v_2^{(2)}(i_1^{(2)})^* p^{(1)} + v_2^{(1)}(i_1^{(1)})^* p^{(2)}}{\Delta v \, \Delta i^*}$$

$$- \tfrac{1}{2} V_{f,2}(x) I^*_{f,1}(x) \frac{v_1^{(2)}(i_2^{(2)})^* p^{(1)} + v_1^{(1)}(i_2^{(1)})^* p^{(2)}}{\Delta v \, \Delta i^*} \qquad (3.23)$$

where $p^{(i)}$ is a shorthand notation for $p^{(i)(i)}$.

3 TWO COUPLED LOSSLESS WAVEGUIDE STRUCTURES

The power propagated by a coupled field transmission line follows from the field voltages $V_{f,p}$ and the currents $I_{f,p}$ of the conductors:

$$P(x) = \tfrac{1}{2}V_{f,1}(x)I_{f,1}^*(x) + \tfrac{1}{2}V_{f,2}(x)I_{f,2}^*(x). \tag{3.24}$$

The equality of (3.23) and (3.24) must be satisfied for any linear combination of the modes. Taking this into account we arrive at four equations:

$$\begin{aligned}
\frac{v_2^{(2)}(i_2^{(2)})^*p^{(1)} + v_2^{(1)}(i_2^{(1)})^*p^{(2)}}{\Delta v \, \Delta i^*} &= 1 \\
\frac{v_1^{(2)}(i_1^{(2)})^*p^{(1)} + v_1^{(1)}(i_1^{(1)})^*p^{(2)}}{\Delta v \, \Delta i^*} &= 1 \\
\frac{v_2^{(2)}(i_1^{(2)})^*p^{(1)} + v_2^{(1)}(i_1^{(1)})^*p^{(2)}}{\Delta v \, \Delta i^*} &= 0 \\
\frac{v_1^{(2)}(i_2^{(2)})^*p^{(1)} + v_1^{(1)}(i_2^{(1)})^*p^{(2)}}{\Delta v \, \Delta i^*} &= 0.
\end{aligned} \tag{3.25}$$

This set of equations can be used to express $v_p^{(i)}$ in terms of $i_p^{(i)}$ or vice versa. The choice of the basic unknowns, either $v_p^{(i)}$ or $i_p^{(i)}$, depends on the choice of the definition of the impedance that will be used in the construction of the circuit model. As we will work out the circuit model for a power–current defined impedance we will use $i_p^{(i)}$ as the basic unknowns. A similar discussion applies for a circuit based upon a power–voltage impedance definition. Solving (3.25) for $v_p^{(i)}$ gives

$$\begin{aligned}
v_1^{(1)} &= \frac{p^{(1)}(i_2^{(2)})^*}{\Delta i^*} \\
v_2^{(1)} &= -\frac{p^{(1)}(i_1^{(2)})^*}{\Delta i^*} \\
v_1^{(2)} &= -\frac{p^{(2)}(i_2^{(1)})^*}{\Delta i^*} \\
v_2^{(2)} &= \frac{p^{(2)}(i_1^{(1)})^*}{\Delta i^*}.
\end{aligned} \tag{3.26}$$

The voltage and the current of a conductor p in an equivalent coupled transmission line are denoted by $V_{e,p}$ and $I_{e,p}$ and are called the equivalent voltage and current of conductor p. The transmission line equations governing the longitudinal variation of $V_{e,p}$ and $I_{e,p}$ follow from the substitution of (3.26) in the transmission line equations (3.14) and (3.15) of the field

3.4 COUPLED EQUIVALENT TRANSMISSION LINES

voltage and current of the conductors. We find:

$$\frac{dV_{e,1}(x)}{dx} = \frac{-j\beta^{(1)}Z_f^{(1)}p^{(1)}i_2^{(2)}(i_2^{(2)})^* - j\beta^{(2)}Z_f^{(2)}p^{(2)}i_2^{(1)}(i_2^{(1)})^*}{|\Delta i|^2} I_{e,1}(x)$$
$$+ \frac{j\beta^{(1)}Z_f^{(1)}p^{(1)}i_1^{(2)}(i_2^{(2)})^* + j\beta^{(2)}Z_f^{(2)}p^{(2)}i_1^{(1)}(i_2^{(1)})^*}{|\Delta i|^2} I_{e,2}(x)$$

$$\frac{dV_{e,2}(x)}{dx} = \frac{j\beta^{(1)}Z_f^{(1)}p^{(1)}i_2^{(2)}(i_1^{(2)})^* + j\beta^{(2)}Z_f^{(2)}p^{(2)}i_2^{(1)}(i_1^{(1)})^*}{|\Delta i|^2} I_{e,1}(x)$$
$$+ \frac{-j\beta^{(1)}Z_f^{(1)}p^{(1)}i_1^{(2)}(i_1^{(2)})^* - j\beta^{(2)}Z_f^{(2)}p^{(2)}i_1^{(1)}(i_1^{(1)})^*}{|\Delta i|^2} I_{e,2}(x)$$

(3.27)

and

$$\frac{dI_{e,1}(x)}{dx} = \frac{-\dfrac{j\beta^{(1)}i_1^{(1)}(i_1^{(1)})^*}{Z_f^{(1)}p^{(1)}} - \dfrac{j\beta^{(2)}i_1^{(2)}(i_1^{(2)})^*}{Z_f^{(2)}p^{(2)}}}{\Delta i^*} V_{e,1}(x)$$
$$+ \frac{-\dfrac{j\beta^{(1)}i_1^{(1)}(i_2^{(1)})^*}{Z_f^{(1)}p^{(1)}} - \dfrac{j\beta^{(2)}i_1^{(2)}(i_2^{(2)})^*}{Z_f^{(2)}p^{(2)}}}{\Delta i^*} V_{e,2}(x)$$

$$\frac{dI_{e,2}(x)}{dx} = \frac{-\dfrac{j\beta^{(1)}i_2^{(1)}(i_1^{(1)})^*}{Z_f^{(1)}p^{(1)}} - \dfrac{j\beta^{(2)}i_2^{(2)}(i_1^{(2)})^*}{Z_f^{(2)}p^{(2)}}}{\Delta i^*} V_{e,1}(x)$$
$$+ \frac{-\dfrac{j\beta^{(1)}i_2^{(1)}(i_2^{(1)})^*}{Z_f^{(1)}p^{(1)}} - \dfrac{j\beta^{(2)}i_2^{(2)}(i_2^{(2)})^*}{Z_f^{(2)}p^{(2)}}}{\Delta i^*} V_{e,2}(x).$$

(3.28)

A general solution to (3.27) and (3.28) is found by substituting the expressions (3.26) for $v_p^{(i)}$ in the general solution (3.18) and (3.19) of the coupled field transmission lines:

$$V_{e,1}(x) = \frac{p^{(1)}(i_2^{(2)})^*}{\Delta i^*}(A^{(1)}e^{-j\beta^{(1)}x} + B^{(1)}e^{j\beta^{(1)}x})$$
$$- \frac{p^{(2)}(i_2^{(1)})^*}{\Delta i^*}(A^{(2)}e^{-j\beta^{(2)}x} + B^{(2)}e^{j\beta^{(2)}x})$$

$$V_{e,2}(x) = -\frac{p^{(1)}(i_1^{(2)})^*}{\Delta i^*}(A^{(1)}e^{-j\beta^{(1)}x} + B^{(1)}e^{j\beta^{(1)}x})$$
$$+ \frac{p^{(2)}(i_1^{(1)})^*}{\Delta i^*}(A^{(2)}e^{-j\beta^{(2)}x} + B^{(2)}e^{j\beta^{(2)}x})$$

(3.29)

and

$$I_{e,1}(x) = \frac{i_1^{(1)}}{Z_f^{(1)}}(A^{(1)}e^{-j\beta^{(1)}x} - B^{(1)}e^{j\beta^{(1)}x}) + \frac{i_1^{(2)}}{Z_f^{(2)}}(A^{(2)}e^{-j\beta^{(2)}x} - B^{(2)}e^{j\beta^{(2)}x})$$

(3.30)

$$I_{e,2}(x) = \frac{i_2^{(1)}}{Z_f^{(1)}}(A^{(1)}e^{-j\beta^{(1)}x} - B^{(1)}e^{j\beta^{(1)}x}) + \frac{i_2^{(2)}}{Z_f^{(2)}}(A^{(2)}e^{-j\beta^{(2)}x} - B^{(2)}e^{j\beta^{(2)}x}).$$

The equivalent line-mode impedance $Z_{lm,e,p}^{(i)}$ of a conductor p associated with a mode i is defined as the ratio of the equivalent voltage to the equivalent current of conductor p contributed by the mode i propagating in the positive x-direction. From the general solution (3.29) and (3.30) we find the following expressions for $Z_{lm,e,p}^{(i)}$:

$$Z_{lm,e,1}^{(1)} = \frac{Z_f^{(1)} p^{(1)} i_2^{(2)*}}{i_1^{(1)} \Delta i^*}$$

$$Z_{lm,e,1}^{(2)} = -\frac{Z_f^{(2)} p^{(2)} i_2^{(1)*}}{i_1^{(2)} \Delta i^*}$$

(3.31)

$$Z_{lm,e,2}^{(1)} = -\frac{Z_f^{(1)} p^{(1)} i_1^{(2)*}}{i_2^{(1)} \Delta i^*}$$

$$Z_{lm,e,2}^{(2)} = \frac{Z_f^{(2)} p^{(2)} i_1^{(1)*}}{i_2^{(2)} \Delta i^*}$$

Back substitution of (3.31) into (3.29) results in a new expression for the equivalent voltages in terms of the current coefficients $i_p^{(i)}$ and the equivalent line-mode impedances $Z_{lm,e,p}^{(i)}$:

$$V_{e,1}(x) = \frac{Z_{lm,e,1}^{(1)} i_1^{(1)}}{Z_f^{(1)}}(A^{(1)}e^{-j\beta^{(1)}x} + B^{(1)}e^{j\beta^{(1)}x})$$

$$+ \frac{Z_{lm,e,1}^{(2)} i_1^{(1)}}{Z_f^{(2)}}(A^{(2)}e^{-j\beta^{(2)}x} + B^{(2)}e^{j\beta^{(2)}x})$$

(3.32)

$$V_{e,2}(x) = \frac{Z_{lm,e,2}^{(1)} i_2^{(1)}}{Z_f^{(1)}}(A^{(1)}e^{-j\beta^{(1)}x} + B^{(1)}e^{j\beta^{(1)}x})$$

$$+ \frac{Z_{lm,e,2}^{(2)} i_2^{(2)}}{Z_f^{(2)}}(A^{(2)}e^{-j\beta^{(2)}x} + B^{(2)}e^{j\beta^{(2)}x}).$$

The electromagnetic field associated with the equivalent coupled field transmission line is given in (3.8) and (3.12) with the voltage coefficients $v_p^{(i)}$ defined in (3.26).

3.5 Circuit model

In this section one of the equivalent coupled transmission lines will be selected as the circuit model for the coupled waveguide structure. As was already mentioned in the introduction we will use the currents flowing on the conductor surfaces as the currents in the circuit model. This choice fixes the circuit model. In 3.5.1 the circuit current representation is given in terms of currents which follow directly from the numerical analysis of the structure. In Sections 3.5.2 and 3.5.3 we derive the circuit line-mode impedances and circuit voltages from the current. In Section 3.5.4 the circuit equations governing the circuit behaviour are presented, and finally, in Section 3.5.5, the power distribution in the circuit is determined and compared to the classical definition introduced by Jansen [3].

3.5.1 Circuit currents

The circuit current $I_{c,p}(x)$ is equal to the total longitudinal current on conductor p. This current follows from the integration of the total magnetic field $\mathbf{H}$ along the circumference $c_{c,p}$ (Fig. 3.1) of conductor p:

$$I_{c,p}(x) = \oint_{c_{c,p}} \mathbf{H} \cdot \mathbf{dl}. \tag{3.33}$$

Substitution of the expression (3.3) for the total magnetic field $\mathbf{H}$ as a linear combination of the two fundamental modes in (3.33) yields the circuit currents $I_{c,p}$ as a linear combination of the field currents $I_f^{(i)}$ of the two modes:

$$I_{c,p}(x) = I_f^{(1)}(x) \oint_{c_{c,p}} \mathbf{H}_t^{(1)} \cdot \mathbf{dl} + I_f^{(2)}(x) \oint_{c_{c,p}} \mathbf{H}_t^{(2)} \cdot \mathbf{dl}. \tag{3.34}$$

The contribution of the longitudinal fields is zero as the integration path lies in the cross-section of the structure.

The current coefficients $i_p^{(i)}$ were introduced to describe the relation (3.7) between the field currents $I_f^{(i)}$ associated with the modes and the field currents $I_{f,p}$ associated with the conductors. In the particular case where the field currents $I_{f,p}$ are the circuit currents we can identify (3.7) with (3.34). This results in the value of the current coefficients $i_p^{(i)}$ associated with our circuit model:

$$i_p^{(i)} = \oint_{c_{c,p}} \mathbf{H}_t^{(i)} \cdot \mathbf{dl}. \tag{3.35}$$

A current coefficient $i_p^{(i)}$ is equal to the integral of the transversal magnetic field $\mathbf{H}_t^{(i)}$ of mode i along the circumference of conductor p.

64   3 TWO COUPLED LOSSLESS WAVEGUIDE STRUCTURES

The wave representation of the circuit currents is given by (3.30) with the values of the current coefficients $i_p^{(i)}$ specified in (3.35). In this representation we can replace the wave coefficients $A^{(i)}$ and $B^{(i)}$ by the total longitudinal currents of the forward and backward propagating modes evaluated in a particular cross-section, for example $x = 0$. If we define $I_p^{(i)+}$ and $I_p^{(i)-}$ as the total longitudinal currents on conductor p in the $x = 0$ cross-section of the waveguide excited by respectively the forward and backward propagating mode i, then the wave representation (3.30) of the conductor currents can be rewritten in the following form:

$$I_{c,1}(x) = I_1^{(1)+} e^{-j\beta^{(1)}x} + I_1^{(1)-} e^{j\beta^{(1)}x} + I_1^{(2)+} e^{-j\beta^{(2)}x} + I_1^{(2)-} e^{j\beta^{(2)}x}$$
$$I_{c,2}(x) = I_2^{(1)+} e^{-j\beta^{(1)}x} + I_2^{(1)-} e^{j\beta^{(1)}x} + I_2^{(2)+} e^{-j\beta^{(2)}x} + I_2^{(2)-} e^{j\beta^{(2)}x}.$$

(3.36)

The total longitudinal currents $I_p^{(i)+}$ and $I_p^{(i)-}$ will be determined by the circuit in which the waveguide is embedded. Once these currents have been determined the circuit currents and voltages are fixed.

The relation between the total longitudinal currents $I_p^{(i)+}$ and the current coefficients $i_p^{(i)}$ is described by the wave coefficients $A^{(i)}$ and $B^{(i)}$ and the impedances $Z_f^{(i)}$ defined in (3.5). Combination of (3.30) and (3.36) gives the following relations for currents flowing in the positive x-direction:

$$I_1^{(1)+} = \frac{i_1^{(1)}}{Z_f^{(1)}} A^{(1)}$$

$$I_1^{(2)+} = \frac{i_1^{(2)}}{Z_f^{(2)}} A^{(2)}$$

$$I_2^{(1)+} = \frac{i_2^{(1)}}{Z_f^{(1)}} A^{(1)}$$

$$I_2^{(2)+} = \frac{i_2^{(2)}}{Z_f^{(2)}} A^{(2)}.$$

(3.37)

Similar expressions hold for the currents associated with the modes propagating in the negative x-direction. From (3.37) we can see that the ratio of the total longitudinal currents of conductors 1 and 2 for each mode i is equal to the ratio of the current coefficients $i_p^{(i)}$ of conductors 1 and 2 for that same mode:

$$\frac{I_1^{(1)\pm}}{I_2^{(1)\pm}} = \frac{i_1^{(1)}}{i_2^{(1)}}$$

$$\frac{I_1^{(2)\pm}}{I_2^{(2)\pm}} = \frac{i_1^{(2)}}{i_2^{(2)}}.$$

(3.38)

3.5 CIRCUIT MODEL

We will use this property in the derivation of an expression for the line-mode impedance $Z^{(i)}_{\text{lm},p}$ in terms of the total longitudinal currents and the power propagated by the modes.

3.5.2 Circuit line-mode impedances

The line-mode impedances $Z^{(i)}_{\text{lm},p}$ associated with a mode i and a conductor p are given by (3.31) with the current coefficients $i^{(i)}_p$ specified in (3.35). The impedances do not depend on the choice of the amplitudes of the fields $\mathbf{E}^{(i)}_t$ and $\mathbf{H}^{(i)}_t$. The proof is based on Table 3.1 which shows how the multiplication of the electric field $\mathbf{E}^{(i)}_t$ by a complex factor $\delta^{(i)}$ affects the other fields, the field voltages and currents, and the current and voltage coefficients.

Table 3.1 Multiplication factors for the fields and the field quantities

$\mathbf{E}^{(i)}_t \to \delta^{(i)} \mathbf{E}^{(i)}_t$	$\mathbf{H}^{(i)}_t \to \dfrac{\mathbf{H}^{(i)}_t}{(\delta^{(i)})^*}$
$\mathbf{H}^{(i)}_l \to \delta^{(i)} \mathbf{H}^{(i)}_l$	$\mathbf{E}^{(i)}_l \to \dfrac{\mathbf{E}^{(i)}_l}{(\delta^{(i)})^*}$
$V^{(i)}_f \to \dfrac{V^{(i)}_f}{\delta^{(i)}}$	$I^{(i)}_f \to I^{(i)}_f (\delta^{(i)})^*$
$v^{(i)}_p \to v^{(i)}_p \delta^{(i)}$	$i^{(i)}_p \to \dfrac{i^{(i)}_p}{(\delta^{(i)})^*}$

In order to construct Table 3.1 we used: the Maxwell equations (2.29) and (2.30) of Chapter 2 to find the transformation of the three fields $\mathbf{E}^{(i)}_t$, $\mathbf{H}^{(i)}_t$, and $\mathbf{H}^{(i)}_l$; the general field representation (3.2) and (3.3) to find the transformation of the field voltage $V^{(i)}_f$ and current $I^{(i)}_f$ of the modes; and (3.26) and (3.35) to find the transformation of the current and voltage coefficients. We took into account that the total field $[\mathbf{E}, \mathbf{H}]$ in (3.2) and (3.3) remains unchanged under this transformation. We leave it as an exercise to the reader to prove, using Table 3.1, that the impedances $Z^{(i)}_{\text{lm},p}$ do not depend on the choice of $\delta^{(i)}$ and therefore have a unique value for each choice of the current coefficients $i^{(i)}_p$ and in particular the choice (3.35) made for the present circuit model.

The expressions (3.31) for the impedances $Z^{(i)}_{\text{lm},p}$ are not convenient for practical use because they contain the quantities $p^{(i)}$, $i^{(i)}_p$, and $Z^{(i)}_f$. These quantities do not follow unequivocally from the numerical analysis of the field problem because they depend on the choice of $\delta^{(i)}$. However, we can rewrite the impedances in terms of variables which are not affected by $\delta^{(i)}$ and are associated with the total modal field.

First, we introduce the power propagated by a mode in the positive x-direction using (3.2), (3.3), and (3.5):

$$P^{(i)+} = \frac{|A^{(i)}|^2}{2(Z_f^{(i)})^*} p^{(i)}. \tag{3.39}$$

This expression for the power $P^{(i)+}$ follows from the integration of the longitudinal component of the Poynting vector over the cross-section S of the waveguide. The wave coefficient $A^{(i)}$ can be replaced by an expression in terms of the current coefficient $i_p^{(i)}$, the total longitudinal current $I_p^{(i)+}$, and the impedance $Z_f^{(i)}$ using (3.37). This leads to

$$P^{(i)+} = \frac{1}{2} \frac{I_1^{(i)+}(I_1^{(i)+})^*}{i_1^{(i)}(i_1^{(i)})^*} Z_f^{(i)} p^{(i)}. \tag{3.40}$$

Using (3.40) we can replace the product of $Z_f^{(i)}$ and $p^{(i)}$ in the impedance expressions (3.31) by an expression containing $P^{(i)+}$ and current coefficients $i_p^{(i)}$. Next, we eliminate the current coefficients $i_p^{(i)}$ by replacing them by the total longitudinal currents $I_p^{(i)+}$ using (3.38). These steps lead to the final expressions for the impedances $Z_{\text{lm},p}^{(i)}$:

$$\begin{aligned}
Z_{\text{lm},1}^{(1)} &= \frac{2P^{(1)+}(I_2^{(2)+})^*}{\Delta I^* I_1^{(1)+}} \\[4pt]
Z_{\text{lm},1}^{(2)} &= -\frac{2P^{(2)+}(I_2^{(1)+})^*}{\Delta I^* I_1^{(2)+}} \\[4pt]
Z_{\text{lm},2}^{(1)} &= -\frac{2P^{(1)+}(I_1^{(2)+})^*}{\Delta I^* I_2^{(1)+}} \\[4pt]
Z_{\text{lm},2}^{(2)} &= \frac{2P^{(2)+}(I_1^{(1)+})^*}{\Delta I^* I_2^{(2)+}}
\end{aligned} \tag{3.41}$$

with

$$\Delta I = I_1^{(1)+} I_2^{(2)+} - I_1^{(2)+} I_2^{(1)+}. \tag{3.42}$$

Equivalent results would have been obtained if we had worked with the modes propagating in the negative x-direction. The above expressions for $Z_{\text{lm},p}^{(i)}$ no longer contain the arbitrary impedances $Z_f^{(i)}$. It is also important to note that the impedances in (3.41) do not depend on an arbitrary multiplication factor involved in the choice of the eigencurrents. For example, in $Z_{\text{lm},p}^{(1)}$ the power $P^{(1)+}$ is proportional to the square of $I_1^{(1)+}$; but the remaining terms on the right-hand side of the expression for $Z_{\text{lm},p}^{(1)}$ are inversely proportional to the square of $I_1^{(1)+}$.

From (3.41) it is easy to prove that the ratio of the line-mode impedance of conductor 1 to the line-mode impedance of conductor 2 is the same for

3.5 CIRCUIT MODEL

both modes. For the first mode this ratio is given by

$$\frac{Z_{lm,1}^{(1)}}{Z_{lm,2}^{(1)}} = -\frac{(I_2^{(2)+})^* I_2^{(1)+}}{I_1^{(1)+}(I_2^{(2)+})^*} \tag{3.43}$$

while for the second mode we have

$$\frac{Z_{lm,1}^{(2)}}{Z_{lm,2}^{(2)}} = -\frac{I_2^{(2)+}(I_2^{(1)+})^*}{(I_1^{(1)+})^* I_2^{(2)+}}. \tag{3.44}$$

If we take the constant phase property of the field $\mathbf{H}_t^{(i)}$ into account we can drop the complex conjugate operators in (3.43) and (3.44). Hence both ratios are equal:

$$\frac{Z_{lm,1}^{(1)}}{Z_{lm,2}^{(1)}} = \frac{Z_{lm,1}^{(2)}}{Z_{lm,2}^{(2)}} = -\frac{I_2^{(2)+} I_2^{(1)+}}{I_1^{(1)+} I_1^{(2)+}}. \tag{3.45}$$

The above expression also shows that the line-mode impedance ratio can be expressed in terms of the total longitudinal currents. Tripathi [4] shows that (3.45) directly results from the requirement that the transmission line equations of the equivalent circuit are reciprocal.

3.5.3 Circuit voltages

Now that we have calculated the circuit currents $I_{c,p}$ and the impedances $Z_{lm,p}^{(i)}$ we can derive an expression for the circuit voltage. In the expression (3.32) for the voltage on the conductors of an equivalent transmission line we can introduce the total longitudinal currents using (3.37). This amounts to:

$$\begin{aligned}
V_{c,1}(x) &= Z_{lm,1}^{(1)}(I_1^{(1)+} e^{-j\beta^{(1)}x} - I_1^{(1)-} e^{j\beta^{(1)}x}) \\
&\quad + Z_{lm,1}^{(2)}(I_1^{(2)+} e^{-j\beta^{(2)}x} - I_1^{(2)-} e^{j\beta^{(2)}x}) \\
V_{c,2}(x) &= Z_{lm,2}^{(1)}(I_2^{(1)+} e^{-j\beta^{(1)}x} - I_2^{(1)-} e^{j\beta^{(1)}x}) \\
&\quad + Z_{lm,p}^{(2)}(I_2^{(2)+} e^{-j\beta^{(2)}x} - I_2^{(2)-} e^{j\beta^{(2)}x}).
\end{aligned} \tag{3.46}$$

The voltages in the circuit model associated with the modes propagating in the positive x-direction can be expressed in terms of the wave coefficients $A^{(i)}$:

$$V_p^{(i)+} = v_p^{(i)} A^{(i)}. \tag{3.47}$$

3.5.4 Circuit equations

The circuit equations take the same form as the coupled equivalent transmission line equations (3.27) and (3.28) but now we have assigned a particular value to the current coefficients and the power impedance

68   3 TWO COUPLED LOSSLESS WAVEGUIDE STRUCTURES

products. The values of the current coefficients (3.35) determine the circuit model in a unique way. It can easily be shown that the coefficients in the circuit equations remain unchanged if we multiply an arbitrary field component of one of the two modes by a complex factor δ, where of course the total field is left unchanged.

Starting from (3.27) and (3.28) and using the relations (3.26), (3.37), and (3.47), the circuit equations can be expressed in terms of impedance and admittance elements per unit of length which are given in terms of voltages and currents associated with the forward propagating modes. Simple manipulations yield:

$$\frac{dV_{c,p}(x)}{dx} = \frac{-j\beta^{(1)}V_p^{(1)+}I_2^{(2)+} + j\beta^{(2)}V_p^{(2)+}I_2^{(1)+}}{\Delta I} I_{c,1}$$

$$+ \frac{j\beta^{(1)}V_p^{(1)+}I_1^{(2)+} - j\beta^{(2)}V_p^{(2)+}I_1^{(1)+}}{\Delta I} I_{c,2} \quad (3.48)$$

and

$$\frac{dI_{c,p}(x)}{dx} = \frac{-j\beta^{(1)}I_p^{(1)+}V_2^{(2)+} + j\beta^{(2)}I_p^{(2)+}V_2^{(1)+}}{\Delta V} V_{c,1}$$

$$+ \frac{j\beta^{(1)}I_p^{(1)+}V_1^{(2)+} - j\beta^{(2)}I_p^{(2)+}V_1^{(1)+}}{\Delta V} V_{c,2}. \quad (3.49)$$

It can be proved that the mutual impedances and the mutual admittances are equal. This reciprocity property is also a consequence of the orthogonality of the two fundamental modes as will be demonstrated in Chapter 4 for the general multiconductor case.

3.5.5 *Power and field distribution in the circuit*

We will now investigate the distribution of the power of each mode over the two conductors. We consider only waves propagating in the positive x-direction; hence the currents in (3.36) flowing in the negative x-direction are equal to zero. From equations (3.36) and (3.46) for the circuit currents and voltages we can calculate the power $P_{c,p}$ propagated by each transmission line p in the circuit model. We find:

$$P_{c,1} = \tfrac{1}{2}Z_{\text{lm},1}^{(1)}|I_1^{(1)+}|^2 + \tfrac{1}{2}Z_{\text{lm},1}^{(2)}|I_1^{(2)+}|^2$$

$$+ \tfrac{1}{2}Z_{\text{lm},1}^{(1)}I_1^{(1)+}(I_1^{(2)+})^* e^{-j(\beta^{(1)}-\beta^{(2)})x}$$

$$+ \tfrac{1}{2}Z_{\text{lm},1}^{(2)}I_1^{(2)+}(I_1^{(1)+})^* e^{-j(\beta^{(2)}-\beta^{(1)})x} \quad (3.50)$$

3.5 CIRCUIT MODEL

and

$$P_{c,2} = \tfrac{1}{2}Z^{(1)}_{lm,2}|I^{(1)+}_2|^2 + \tfrac{1}{2}Z^{(2)}_{lm,2}|I^{(2)+}_2|^2$$
$$+ \tfrac{1}{2}Z^{(1)}_{lm,2}I^{(1)+}_2(I^{(2)+}_2)^* e^{-j(\beta^{(1)}-\beta^{(2)})x}$$
$$+ \tfrac{1}{2}Z^{(2)}_{lm,2}I^{(2)+}_2(I^{(1)+}_2)^* e^{-j(\beta^{(2)}-\beta^{(1)})x}. \qquad (3.51)$$

In each of the above expressions the first term represents the contribution from the first mode, the second term from the second mode, while the last two terms represent the power coupling between the modes. Defining $P^{(i)+}_p$ as the part of the power associated with mode i and transmitted by conductor p in the absence of the other mode, we find the following expressions for $P^{(i)+}_p$ from (3.50) and (3.51):

$$P^{(i)+}_p = \tfrac{1}{2}Z^{(i)}_{lm,p}|I^{(i)+}_p|^2. \qquad (3.52)$$

The total power $P^{(i)+}$ associated with each mode i in the absence of the other mode is given by

$$P^{(i)+} = P^{(i)+}_1 + P^{(i)+}_2. \qquad (3.53)$$

The total power P propagated by the coupled waveguide is the sum of the power propagated by the two conductors. If we take the impedance current relations (3.41) into account we find that the coupling terms in (3.50) and (3.51) cancel out in the total power budget. The total propagated power is also given as the sum of the powers associated with the modes:

$$P = P_{c,1} + P_{c,2} = P^{(1)+} + P^{(2)+}. \qquad (3.54)$$

The fact that the power coupling terms in (3.50) and (3.51) cancel out in the total power budget is a consequence of the power orthogonality of the modes.

The power terms in (3.52) can be expressed in terms of currents and the power propagated by the modes by replacing the impedances in (3.52) by their expressions (3.41). We find:

$$P^{(1)+}_1 = \frac{P^{(1)+}I^{(2)+}_2 I^{(1)+}_1}{\Delta I}$$

$$P^{(2)+}_1 = \frac{P^{(2)+}I^{(1)+}_2 I^{(2)+}_1}{\Delta I}$$

$$P^{(1)+}_2 = -\frac{P^{(1)+}I^{(2)+}_1 I^{(1)+}_2}{\Delta I}$$

$$P^{(2)+}_2 = \frac{P^{(2)+}I^{(1)+}_1 I^{(2)+}_2}{\Delta I}. \qquad (3.55)$$

The phase property of the field was used to drop the complex conjugate operators. In (3.55) we have now expressed the power propagated by conductor p ($p = 1, 2$) associated with mode i ($i = 1, 2$) in terms of the total power propagated by each mode separately. The expressions (3.55) lead to the following relation between the powers $P_p^{(i)+}$:

$$\frac{P_1^{(1)+}}{P_2^{(1)+}} = \frac{P_2^{(2)+}}{P_1^{(2)+}}. \tag{3.56}$$

This relation shows that, for example, if 30 per cent of the power of mode 1 is propagated by line 1 and 70 per cent by line 2, it automatically follows that 70 per cent of the power of mode 2 is propagated by line 1 and 30 per cent by line 2. This reasoning concerns the individual modes. If both modes are present, part of the power oscillates between the two lines as a function of the longitudinal distance along the waveguide structure. The typical length involved is the coupling length which depends on the difference of the propagation constants of the modes as we can see from (3.50) and (3.51). These equations contain phase factors in the third and fourth term which depend on this difference.

In order to find the electric and magnetic fields associated with each conductor in the circuit model we replace in (3.8) and (3.12) the field voltages and currents associated with the conductors by the circuit voltages and currents. The total field is now expressed in terms of the circuit voltages and currents:

$$\mathbf{E} = \mathbf{E}_{t,1} V_{c,1} + \mathbf{E}_{t,2} V_{c,2} + R_0 \mathbf{E}_{l,1} I_{c,1} + R_0 \mathbf{E}_{l,2} I_{c,2} \tag{3.57}$$

and

$$\mathbf{H} = \mathbf{H}_{t,1} I_{c,1} + \mathbf{H}_{t,2} I_{c,2} + \frac{\mathbf{H}_{l,1}}{R_0} V_{c,1} + \frac{\mathbf{H}_{l,2}}{R_0} V_{c,2}. \tag{3.58}$$

The field components in (3.57) and (3.58) are given in (3.9) and (3.13) with the voltage coefficients given in (3.26) and the current coefficients in (3.35). The field expressions (3.57) and (3.58) will now be used to find the fields which build up the power $P_p^{(i)+}$. This power is described in terms of the part of the electric and magnetic field of mode i which is propagated by conductor p:

$$P_p^{(i)+} = \frac{1}{2} \int_S \int [\mathbf{E}_p^{(i)+} \times (\mathbf{H}_p^{(i)+})^*] \cdot \mathbf{1}_x \, dS. \tag{3.59}$$

The electric field $\mathbf{E}_p^{(i)+}$ and the magnetic field $\mathbf{H}_p^{(i)+}$ follow from (3.57) and (3.58). Those fields can be expressed in terms of quantities which follow directly from a numerical simulation. Combinations of the field representations (3.9), (3.13), (3.57), and (3.58) with the expressions (3.37) and (3.47) for the

3.5 CIRCUIT MODEL

circuit currents and voltages yields the following expressions for the fields associated with conductor p and propagated by mode i:

$$\mathbf{E}_1^{(i)+} = Z_{\text{lm},1}^{(i)} \frac{V_2^{(2)+}\mathbf{E}_{\text{tot},t}^{(1)+} - V_2^{(1)+}\mathbf{E}_{\text{tot},t}^{(2)+}}{\Delta V} I_1^{(i)+}$$

$$\mathbf{H}_1^{(i)+} = \frac{I_2^{(2)+}\mathbf{H}_{\text{tot},t}^{(1)+} - I_2^{(1)+}\mathbf{H}_{\text{tot},t}^{(2)+}}{\Delta I} I_1^{(i)+}$$

(3.60)

for conductor 1 and

$$\mathbf{E}_2^{(i)+} = Z_{\text{lm},2}^{(i)} \frac{-V_1^{(2)+}\mathbf{E}_{\text{tot},t}^{(1)+} + V_1^{(1)+}\mathbf{E}_{\text{tot},t}^{(2)+}}{\Delta V} I_2^{(i)+}$$

$$\mathbf{H}_2^{(i)+} = \frac{-I_1^{(2)+}\mathbf{H}_{\text{tot},t}^{(1)+} + I_1^{(1)+}\mathbf{H}_{\text{tot},t}^{(2)+}}{\Delta I} I_2^{(i)+}$$

(3.61)

for conductor 2. The subscript 'tot' denotes that the field component is a total field associated with a mode. For a certain conductor, the fields associated with the two modes are proportional and differ only in their amplitude which is determined by the current and the line-mode impedance associated with the conductor and the mode. The fields associated with each conductor p are an average of the two modal fields. For example, the magnetic field associated with conductor 1 is such that it produces a total longitudinal current of 1 ampere on conductor 1 and a current on conductor 2 of which the total longitudinal component is zero. However, we stress that the current on conductor 2 is not identically zero.

In the past Jansen [3] proposed the following expression for the partial powers $P_p^{(i)+}$:

$$P_p^{(i)+} = \frac{1}{2} \int_S \int [\mathbf{E}_{\text{tot}}^{(i)+} \times (\mathbf{H}_p^{(i)+})^*] \cdot \mathbf{1}_x \, dS. \quad (3.62)$$

In (3.62) the electric field $\mathbf{E}_{\text{tot}}^{(i)+}$ represents the total field propagated by mode i in the positive x-direction. The magnetic field $\mathbf{H}_p^{(i)+}$ is now defined as the magnetic field due to the current on conductor p excited by the mode j. The current on the other conductor is identically zero. So the magnetic field $\mathbf{H}_p^{(i)+}$ in (3.62) differs from the magnetic field $\mathbf{H}_p^{(i)+}$ associated with our circuit model. It can be shown that the value for the power in (3.62) gives the correct result in the quasi-TEM limit. The expression (3.62), which is not part of a complete circuit model for coupled lines, seems to be inspired by a closely related one:

$$P_p^{(i)+} = \frac{1}{2} \int_S \int [\mathbf{E}_p^{(i)+} \times (\mathbf{H}_p^{(i)+})^*] \cdot \mathbf{1}_x \, dS. \quad (3.63)$$

3 TWO COUPLED LOSSLESS WAVEGUIDE STRUCTURES

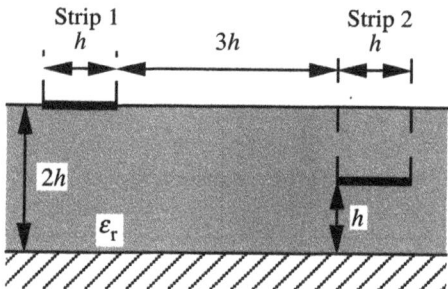

Fig. 3.2. Cross-section of the analysed microstrip configuration.

Now the electric field $\mathbf{E}_p^{(i)+}$ is defined as the electric field for mode i due to the voltage on conductor i, while the voltage on the other conductor is zero. In the quasi-static limit (3.62) and (3.63) are equivalent. The expression (3.63) is the classical definition of the power distribution in TEM waveguides. However, since a typical full-wave analysis using the spectral domain method yields total electric fields and partial magnetic fields, (3.62) is an obvious extension of (3.63).

Line-mode impedances derived with the power definition (3.62) of Jansen have the same value in the quasi-static limit as the impedances derived from the power distribution (3.55). However, for higher frequencies both power distributions yield different results. The power definition presented in this chapter follows from a consistent circuit model. Jansen's definition is only an empirical extension of the exact quasi-static power distribution.

We conclude this section with an illustration of the difference between the power definition of Jansen and the power distribution found in this chapter. Figure 3.2 shows the coupled microstrip configuration under study. The value of h is 1 mm and the relative permittivity ε_r of the substrate is 4. Using the full-wave analysis technique presented in Chapter 7 the power distribution can be calculated according to the definition proposed by Jansen and to the expression found in the circuit model presented in this chapter. Figure 3.3 shows the four line-mode impedances.

Mode 1 corresponds to the quasi-odd mode while mode 2 corresponds to the quasi-even mode. The solid lines in Fig. 3.3 represent the impedances found with the power distribution proposed in this chapter. The dashed lines give the results which follow from Jansen's definition (3.62). The differences between Jansen's approach and ours increase as a function of the frequency. For $Z_{\text{lm},2}^{(1)}$ and $Z_{\text{lm},1}^{(2)}$ the difference remains small. As a check on relation (3.45), which is a consequence of the reciprocity, we calculated the following variable:

$$\xi = \left| 1 - \frac{Z_{\text{lm},1}^{(1)} Z_{\text{lm},2}^{(2)}}{Z_{\text{lm},1}^{(2)} Z_{\text{lm},2}^{(1)}} \right| \tag{3.64}$$

3.5 CIRCUIT MODEL

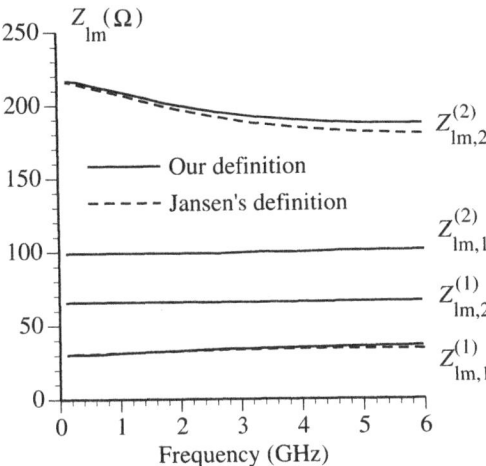

Fig. 3.3. Line-mode impedances as a function of frequency for the configuration of Fig. 3.2.

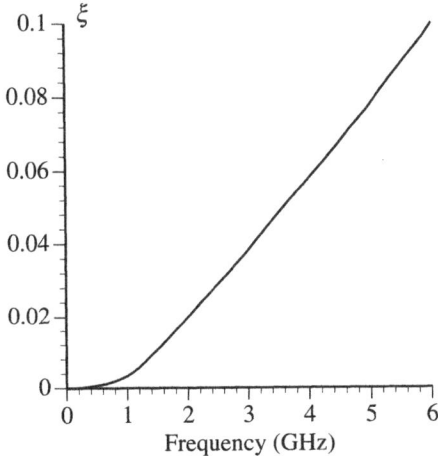

Fig. 3.4. ξ as a function of the frequency using the power distribution of Jansen [3] applied to the configuration of Fig. 3.2.

which must be zero. Figure 3.4 shows ξ as a function of the frequency using Jansen's definition. This figure clearly shows that the approach based on (3.62) yields a correct circuit model only in the quasi-TEM limit. In our approach ξ is of course zero. The fact that reciprocity is violated if one uses (3.62) as the basis for a circuit model is also discussed in [5]. As explained in Section 1.2.8 and discussed in more detail in the next chapter, a consistent circuit model as given in the present chapter leading to (3.45) and (3.55), yields a reciprocal circuit model. This reciprocity follows automatically from the power orthogonality of the modes in the lossless waveguide structure.

References

[1] Faché, N. and De Zutter, D. (1990). New high frequency circuit model for coupled lossless and lossy waveguide structures. *IEEE Transactions on Microwave Theory and Techniques*, **MTT-38**, 3, 252–259.
[2] Collin, R. E. (1960). *Field Theory of Waves*. McGraw-Hill, New York.
[3] Jansen, R. H. (1979). Unified user-oriented computation of shielded, covered and open planar microwave and millimeter-wave transmission-line characteristics. *Microwaves, Optics and Acoustics*, **3**, 1, 14–22.
[4] Tripathi, V. K. (1975). Asymmetric coupled transmission lines in an inhomogeneous medium. *IEEE Transactions on Microwave Theory and Techniques*, **MTT-23**, 9, 734–739.
[5] Wiemer, L. and Jansen, R. H. (1986). Reciprocity related definition of strip characteristic impedance for multiconductor hybrid-mode transmission lines. *Microwave and Optical Technology Letters*, **1**, 22–25.

4
CIRCUIT DESCRIPTION OF GENERAL LOSSY MULTICONDUCTOR WAVEGUIDE STRUCTURES

4.1 Introduction

In the previous chapters the network description of single and coupled multiconductor waveguides was developed in a systematic way highlighting the physical background and the concepts introduced in order to come to a systematic network description.

In Chapter 3 the network description for coupled waveguides was restricted to two coupled lossless waveguides. In the present chapter the network description will be extended to an arbitrary number of coupled multiconductor waveguides and losses will be included as well. As we assume that the reader is familiar with the concepts introduced in the previous chapters we will restrict ourselves to the general outline and the final results of the problem. However, the chapter, and in particular the notation used to indicate field quantities on the one hand and circuit quantities on the other hand, is organized in such a way that the reader should be able to read this chapter independently of the previous ones.

In contradistinction to the previous chapters we will use a matrix formalism throughout this chapter. This guarantees a compact description which was deliberately avoided in the previous chapter to clarify the proposed concepts. Another particular advantage of using a matrix formalism is that it lends itself much more easily to implementation on a computer.

The discussion in Section 4.3 starts with the generalization of the modal field representation and the construction of a set of field transmission lines to describe these modal fields. In Section 4.4 the modal description is transformed into a circuit description either by means of the power–current or by the power–voltage formulation. In Section 4.5 the final transmission line model is introduced and the generalized frequency-dependent capacitance, inductance, conductance, and resistance matrices are defined. We also compare the characteristic impedance matrix, which describes the relation between the circuit voltage and the circuit currents, with the line-mode impedance matrix, which describes the relation between the circuit voltage and current of one mode and associated with each conductor. In the last part of Section 4.5 we prove that, although the waveguide is a reciprocal electromagnetic structure, the derived circuit model is not necessarily

76   4 LOSSY MULTICONDUCTOR WAVEGUIDE STRUCTURES

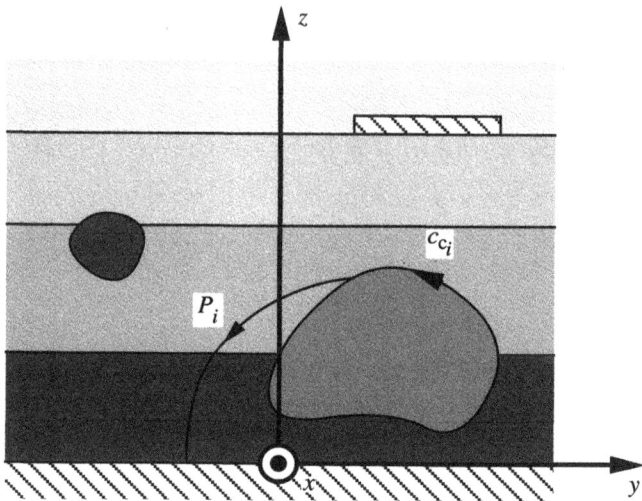

Fig. 4.1. Geometry of the cross-section of a general coupled waveguide structure.

reciprocal. In the last section of this chapter we present a case study for two coupled lossy asymmetrical microstrip lines. Publications specifically concerning the work in this chapter can be found in [1], [2], [3], and [4].

4.2 General coupled waveguide structure

Figure 4.1 shows the cross-section of a general coupled lossy waveguide structure consisting of $N + 1$ conductors in a lossless or lossy inhomogeneous dielectric. Each conductor has an arbitrary cross-section and is uniform along its length (longitudinal x-direction). The conductors can either be perfectly conducting or exhibit a finite conductivity. The $(N + 1)$th line is chosen as the reference or ground conductor. In most cases of practical importance this will be a ground plane at the bottom or the top of the structure or a shielding surrounding all other conductors.

We focus our attention on the N fundamental modes of the coupled multiconductor structure.

4.3 Modal field representation

The total electromagnetic field consists of the sum of the partial fields of the N fundamental modes of the structure under study:

$$\mathbf{E}(x, y, z) = \mathbf{V}_f^T(x) \mathbf{E}_t^m(y, z) + R_0 \mathbf{I}_f^T(x) \mathbf{E}_l^m(y, z)$$
$$\mathbf{H}(x, y, z) = \mathbf{I}_f^T(x) \mathbf{H}_t^m(y, z) + \frac{1}{R_0} \mathbf{V}_f^T(x) \mathbf{H}_l^m(y, z).$$
(4.1)

4.3 MODAL FIELD REPRESENTATION

We have grouped the N field voltages $V_{f,i}$ ($i = 1, 2, \ldots, N$) into the column vector V_f and the corresponding field currents $I_{f,i}$ into the column vector I_f. Throughout this chapter, the bold italic quantities are matrices or, as a special case of that, a column vector. Each time we introduce a new matrix, we will carefully indicate its dimensions. The superscript 'T' is the transposition operator. The subscript 'f' indicates a field quantity.

The column vector E_t^m consists of N elements $\mathbf{E}_{t,i}^m$ ($i = 1, 2, \ldots, N$). The superscript 'm' reminds the reader of the modal character of the quantity. In itself, each element of this column vector is a field vector quantity representing the modal transversal electric field associated with mode i.

Analogous interpretations can be given for the other modal column vectors. In the following we will refer to E_t^m as the transversal electric field. Although this quantity is actually a conglomerate of the transversal electric fields of the N modes, this shorthand description is preferred for conciseness. The dimension of these fields is m^{-1}. The transversal (subscript 't') and longitudinal (subscript 'l') fields depend only on the transversal space coordinates y and z. $R_0 = \sqrt{\mu_0/\varepsilon_0}$ is the characteristic impedance of free space.

We can repeat the method of calculation of Section 2.2.2 for each mode individually. Starting from Maxwell's equations and (4.1), it is then easy to show that the field voltages and currents are related by the following transmission line equations:

$$-\frac{d}{dx} V_f(x) = j\beta Z_f I_f(x)$$

$$-\frac{d}{dx} I_f(x) = j\beta Z_f^{-1} V_f(x). \tag{4.2}$$

Here β is a $N \times N$ matrix but only the diagonal elements β_{ii} ($i = 1, 2, \ldots, N$) differ from zero with $\beta_{ii} = \beta_i$, where β_i is the complex propagation constant of the ith mode. The matrix Z_f is also a diagonal matrix: $Z_{f,ij} = 0$ for $i \neq j$ and $Z_{f,ii} = Z_{f,i}$, the complex modal field impedance belonging to mode i. The superscript '-1' stands for the inverse matrix operator. The eigenvalue equations (2.31) and (2.32) remain valid. The solution of one of these equations yields the modal field distributions and the corresponding propagation constants.

Solution of (4.2) leads to

$$V_f(x) = A e^{-j\beta x} + B e^{j\beta x}$$
$$= V_f^+(x) + V_f^-(x) \tag{4.3}$$

and

$$I_f(x) = Z_f^{-1}[V_f^+(x) - V_f^-(x)]$$
$$= I_f^+(x) + I_f^-(x). \quad (4.4)$$

The column vectors A and B (dimension N) are constants determined by the boundary conditions at the beginning ($x = 0$) and at the end ($x = L$) of the coupled waveguide structure. We have used the superscripts '+' and '−' to distinguish between waves propagating in the positive and negative x-direction.

4.4 Transformation from a modal to a circuit description

4.4.1 *General formulation*

In order to simulate the hybrid waveguide structure under study with a circuit simulator, we have to transform the modal description into a circuit model consisting of coupled lossy dispersive transmission lines. For TEM structures, the conductor voltages and currents could be calculated in an unambiguous way by line integrals of the electric and magnetic fields. For non-TEM structures however, there is no such unique definition of conductor voltage and current. In this chapter we use the power–current formulation established in Chapter 2. Thus, the circuit current belonging to conductor i is chosen to be identical to the total longitudinal current flowing along that conductor. Furthermore, both the circuit model representation and the real waveguide structure are required to have the same complex propagation constants and to propagate the same average complex power.

The discussion proceeds in three steps: first the circuit current is defined followed in a second step by the circuit voltage. The arbitrariness introduced in the first two steps is removed by the introduction of the power conservation principle in a third step. Unlike the development in Chapters 2 and 3 we do not introduce the power principle first and hence we do not invoke the equivalent transmission line description.

4.4.2 *Circuit current*

As stated above, the circuit current is defined as the total longitudinal current flowing along conductor i:

$$I_{c,i}(x) = \oint_{c_{c_i}} \mathbf{H}(x, y, z) \cdot \mathbf{dl} \quad (4.5)$$

4.4 TRANSFORMATION FROM A MODAL DESCRIPTION

and substituting from (4.1):

$$I_{c,i}(x) = \oint_{c_{c_i}} I_f^T(x) H_t^m(y, z) \cdot dl \tag{4.6}$$

with $i = 1, 2, \ldots, N$. The current is a clear physical quantity that consists of contributions of the N propagating modes. The integration extends over the boundary c_{c_i} (see Fig. 4.1) of the cross-section of conductor i. Equation (4.6) can be written in a compact matrix form as

$$I_c(x) = M_I I_f(x)$$
$$M_{I,ij} = \oint_{c_{c_i}} H_{t,j}^m(y, z) \cdot dl. \tag{4.7}$$

We have introduced the column vector I_c formed by the N circuit currents $I_{c,i}$ ($i = 1, 2, \ldots, N$). Here and in the following the subscript 'c' is used to distinguish between circuit quantities and field quantities (subscript 'f').

M_I is the frequency-dependent transformation matrix between modal currents and circuit currents. Each element $M_{I,ij}$ ($i, j = 1, 2, \ldots, N$) expresses the contribution of the field current $I_{f,j}$ of mode j to the circuit current $I_{c,i}$ on conductor i. Observe that all $M_{I,ij}$s with $i = 1, 2, \ldots, N$ but for the same j value, that is, each column of M_I, can be multiplied by an arbitrary complex factor δ_j if $I_{f,j}$ is divided by that same factor. Consequently, the transformation matrix M_I is not uniquely defined. From the physics of the problem it is clear that the ratio of the contribution to the circuit current on conductor i from a particular mode p, to the contribution to the circuit current on conductor j by that same mode, is unequivocally determined, that is, the ratios $M_{I,ip}/M_{I,jp}$ ($i, j, p = 1, 2, \ldots, N$) are fixed.

4.4.3 Circuit voltage

The circuit voltage column vector $V_c(x)$ has still to be determined. Now we will look for a practical representation suited to further matrix calculations. Later on, we will use the power–current formulation to define the remaining unknown parameters.

By analogy with (4.7) we represent the circuit voltage vector V_c as a superposition of the modal voltages V_f introduced in (4.1):

$$V_c(x) = M_V V_f(x) \tag{4.8}$$

where M_V is the frequency-dependent $N \times N$ transformation matrix between modal voltages and circuit voltages.

4.4.4 Power conservation

We will now assign a value to the arbitrary matrices M_V by looking at the propagated power. The circuit model and the coupled waveguide structure under study should propagate the same complex power. The average complex power propagated by the coupled hybrid waveguide structure can be found by integrating the longitudinal component of Poynting's vector over the total cross-section S of the structure. This leads to

$$P_f(x) = \tfrac{1}{2} V_f^T(x) p I_f^*(x). \tag{4.9}$$

The elements of p are given by

$$p_{ij} = \int\int_S [\mathbf{E}_{t,i}^m \times (\mathbf{H}_{t,j}^m)^*] \cdot \mathbf{1}_x \, dS \tag{4.10}$$

with $i, j = 1, 2, \ldots, N$. The superscript '*' represents the complex conjugate operator. p_{ij} is the average power propagated in the x-direction by the electric field of mode i and the magnetic field of mode j.

On the other hand, the average complex power P_c propagated by the coupled transmission line model is given by

$$P_c(x) = \tfrac{1}{2} V_c^T(x) I_c^*(x), \tag{4.11}$$

or, equivalently (using (4.7) and (4.8)),

$$P_c(x) = \tfrac{1}{2} V_f^T(x) M_V^T M_I^* I_f^*(x). \tag{4.12}$$

If P_f and P_c must be equal, (4.9) and (4.12) lead to the requirement:

$$p = M_V^T M_I^* \tag{4.13}$$

which shows that the transformation matrices between circuit currents and field currents on the one hand, and between circuit voltages and field voltages on the other, are not independent. Indeed, from (4.13) we can determine M_V:

$$M_V = (M_I^{*T})^{-1} p^T. \tag{4.14}$$

4.4.5 Arbitrariness of M_I and M_V

As noted at the end of Section 4.2, M_I is not defined in a unique way. This is, for analogous reasons, also the case for M_V. By exploiting the arbitrariness in the definition of I_f and V_f the following substitution is always possible:

$$\begin{aligned} M_I &\to M_I \delta_I \\ M_V &\to M_V \delta_V \end{aligned} \tag{4.15}$$

4.4 TRANSFORMATION FROM A MODAL DESCRIPTION

where δ_I and δ_V are $N \times N$ diagonal matrices. This means, using (4.7) and (4.8), that

$$I_f \to (\delta_I)^{-1} I_f$$
$$V_f \to (\delta_V)^{-1} V_f. \tag{4.16}$$

Equation (4.1) then results in the following transformation rules for the transversal and longitudinal modal field components:

$$E_t^m \to \delta_V E_t^m \qquad E_l^m \to \delta_I E_l^m$$
$$H_t^m \to \delta_I H_t^m \qquad H_l^m \to \delta_V H_l^m. \tag{4.17}$$

From equation (4.2) on the other hand it follows that Z_f transforms as

$$Z_f \to (\delta_V)^{-1} \delta_I Z_f, \tag{4.18}$$

and from (4.10) and (4.13) we have that

$$p \to \delta_V p \delta_I^*. \tag{4.19}$$

This arbitrariness in M_I and M_V, which has no effect on the circuit model as will become clear below, can be removed by demanding, for example, that

$$\int_S \int |E_{t,i}^m|^2 \, dS = \int_S \int |H_{t,i}^m|^2 \, dS = 1 \tag{4.20}$$

with $i = 1, 2, \ldots, N$. This fixes only the amplitude of the elements of δ_I and δ_V. The phase of these elements can be fixed by demanding, for example, that $\mathbf{E}_{t,i}^m(y = 0, z = 0) \cdot \mathbf{1}_z$ and $\mathbf{H}_{t,i}^m(y = 0, z = 0) \cdot \mathbf{1}_z$ are real quantities.

For lossless structures we know from (2.31) and (2.32) that $\mathbf{E}_{t,i}^m(y, z)$ and $\mathbf{H}_{t,i}^m(y, z)$ have a constant phase over the cross-section of the structure. Hence, in this case we can demand that $\mathbf{E}_{t,i}^m(y, z)$ and $\mathbf{H}_{t,i}^m(y, z)$ ($i = 1, 2, \ldots, N$) are real quantities.

4.4.6 Power–voltage circuit model

The power–current model discussed above can readily be adapted to the power–voltage model. This last model is believed to be more suited to describing slotline and coplanar waveguide structures. We refer the reader to Chapters 1 and 2 for a discussion on the choice of the most suitable circuit model.

In the power–voltage model we start with the definition of the circuit voltage. This circuit voltage is defined as the line integral of the total electric field along a suitable path P_i (see Fig. 4.1) between the reference or ground conductor and conductor i:

$$V_{c,i}(x) = \int_{P_i} \mathbf{E}(x, y, z) \cdot \mathbf{dl}, \tag{4.21}$$

4 LOSSY MULTICONDUCTOR WAVEGUIDE STRUCTURES

and substituting from (4.1):

$$V_{c,i}(x) = \int_{P_i} V_f^T(x) E_t^m(y, z) \cdot dl \quad (4.22)$$

with $i = 1, 2, \ldots, N$. Equation (4.22) can be written in a compact matrix form as

$$V_c(x) = M_V V_f(x). \quad (4.23)$$

This defines the transformation matrix M_V. Each element $M_{V,ij}$ ($i,j = 1, 2, \ldots, N$) has now a physical meaning as the line integral of the modal electric field $E_{t,j}^m$ along the path P_i to conductor i.

In the power–voltage formulation the matrix M_I has no physical meaning and is determined from the power condition (4.13):

$$M_I = (M_V^{*T})^{-1} p^*. \quad (4.24)$$

4.5 Coupled transmission line model

In this section the final transmission line model is introduced. The generalized line parameter matrices $R(\omega)$, $G(\omega)$, $L(\omega)$, and $C(\omega)$ are defined and it is shown how these matrices can be derived from the modal fields. Subsequent sections discuss the properties of the circuit model which can be seen as an extension of the more familiar quasi-TEM approximation.

4.5.1 Generalized telegrapher's equations

The generalized telegrapher's equations which govern the circuit representation can be found by substitution of the circuit currents (4.7) and the circuit voltages (4.8) into the modal transmission line equations (4.2):

$$-\frac{d}{dx} V_c(x) = z_c I_c(x)$$
$$-\frac{d}{dx} I_c(x) = y_c V_c(x). \quad (4.25)$$

The $N \times N$ circuit impedance and admittance matrices z_c and y_c are given by

$$z_c = M_V j\beta Z_f M_I^{-1}$$
$$y_c = M_I j\beta Z_f^{-1} M_V^{-1}. \quad (4.26)$$

4.5 COUPLED TRANSMISSION LINE MODEL

Equations (4.25) also lead to the following wave equations:

$$\frac{d^2}{dx^2} V_c(x) = z_c y_c V_c(x)$$
$$= -\Lambda_V^2 V_c(x) \qquad (4.27)$$

$$\frac{d^2}{dx^2} I_c(x) = y_c z_c I_c(x)$$
$$= -\Lambda_I^2 I_c(x)$$

with

$$\Lambda_V^2 = M_V \beta^2 M_V^{-1}$$
$$\Lambda_I^2 = M_I \beta^2 M_I^{-1} \qquad (4.28)$$

and, because β is diagonal

$$\Lambda_V = M_V \beta M_V^{-1}$$
$$\Lambda_I = M_I \beta M_I^{-1}. \qquad (4.29)$$

The (frequency-dependent!) equations (4.25) are very similar to the well-known quasi-static transmission line equations discussed in [5] and [6]. Note however that no static extrapolations or assumptions were made. The non-symmetric matrices $\Lambda_V(\omega)$ and $\Lambda_I(\omega)$ represent the complex voltage and current propagation matrices, respectively. Observe that the substitutions (4.15) do not affect Λ_V and Λ_I. These circuit quantities are indeed defined in a unique way.

Generally speaking, the z_c and y_c matrices are frequency dependent, and we can split them into a real and an imaginary part:

$$z_c(\omega) = R(\omega) + j\omega L(\omega)$$
$$y_c(\omega) = G(\omega) + j\omega C(\omega) \qquad (4.30)$$

where $R(\omega)$, $G(\omega)$, $L(\omega)$, and $C(\omega)$ are the generalized $N \times N$ resistance, capacitance, conductance, and inductance matrices. The circuit model formed by (4.25) and (4.30) is fully compatible with, and is an extension towards, higher frequencies of the well-known TEM and quasi-TEM circuit modes. Some typical static concepts, such as capacitance and inductance, are generalized by (4.30) and introduced in the high-frequency model. Note that at each discrete frequency the model can be seen as a quasi-static model. The resemblance to the quasi-static model is the main reason why the proposed frequency-dependent model is perfectly suited for CAD calculations.

In [7] integral expressions, containing the modal fields, were obtained for the R, G, L, and C circuit parameters of a single waveguide. In this section

84   4 LOSSY MULTICONDUCTOR WAVEGUIDE STRUCTURES

we will extend these to integral expressions for the R, G, L, and C matrices of coupled waveguides. The expressions were first proposed in [4].

We start by substituting (4.10) into the power condition (4.13):

$$M_V^T M_I^* = \int_S \int [E_t^m(y, z) \times H_t^{m*T}(y, z)] \cdot 1_x \, dS. \qquad (4.31)$$

With elementary vector analysis this is rewritten as

$$M_V^T M_I^* = \int_S \int [1_x \times E_t^m(y, z)] \cdot H_t^{m*T}(y, z) \, dS. \qquad (4.32)$$

From the generalized matrix form of the second equation of (2.29) the following expression is obtained:

$$1_x \times E_t^m(y, z) = -j\beta^{-1} Z_f^{-1} [R_0 \nabla_t \times E_t^m(y, z) + j\omega\mu(y, z) H_t^m(y, z)] \qquad (4.33)$$

and from the generalized third equation of (2.29) we obtain

$$R_0 \nabla_t \times E_t^m(y, z) = \frac{1}{j\omega} \nabla_t \times \left[\frac{1}{\varepsilon(y, z)} \nabla_t \times H_t^m(y, z) \right]. \qquad (4.34)$$

Substitution of (4.33) and (4.34) in (4.32) and partial integration yields

$$j\beta Z_f M_V^T M_I^* = \int_S \int \left\{ \frac{1}{j\omega\varepsilon(y, z)} [\nabla_t \times H_t^m(y, z)] \cdot [\nabla_t \times H_t^{m*T}(y, z)] \right. $$
$$\left. + j\omega\mu(y, z) H_t^m(y, z) \cdot H_t^{m*T}(y, z) \right\} dS. \qquad (4.35)$$

In (4.35) it is assumed that the electromagnetic fields satisfy Sommerfeld's radiation condition at infinity or vanish at the boundaries of the waveguide structure. If we again use the third equation of (2.29) and if we pre-multiply by $(M_I^T)^{-1}$ and post-multiply by $(M_I^*)^{-1}$ then (4.35) reduces to

$$(M_I^T)^{-1} j\beta Z_f M_V^T = j\omega \int_S \int [-\varepsilon^*(y, z) R_0^2 E_t^c(y, z) \cdot E_t^{c*T}(y, c)$$
$$+ \mu(y, z) H_t^c(y, c) \cdot H_t^{c*T}(y, z)] \, dS \qquad (4.36)$$

with

$$E_t^c(y, z) = (M_I^T)^{-1} E_t^m(y, z)$$
$$H_t^c(y, z) = (M_I^T)^{-1} H_t^m(y, z). \qquad (4.37)$$

4.5 COUPLED TRANSMISSION LINE MODEL

The left-hand side of (4.36) is equal to z_c^T, hence

$$z_c = R(\omega) + j\omega C(\omega) = j\omega \int\!\!\int_S [-\varepsilon^*(y,z) R_0^2 E_1^{c*}(y,z) \cdot E_1^{cT}(y,z)$$
$$+ \mu(y,z) H_t^{c*}(y,z) \cdot H_t^{cT}(y,z)]\, dS. \quad (4.38)$$

If one starts from

$$M_V^T M_I^* = -\int\!\!\int_S E_t^m(y,z) \cdot [1_z \times H_t^{m*T}(y,z)]\, dS \quad (4.39)$$

instead of equation (4.32) one obtains, in an analogous way, the following expression for $y_c = G + j\omega L$:

$$y_c = G(\omega) + j\omega L(\omega) = j\omega \int\!\!\int_S [\varepsilon(y,z) E_t^{c*}(y,z) \cdot E_t^{cT}(y,z)$$
$$- \frac{\mu^*(y,z)}{R_0^2} H_1^{c*}(y,z) \cdot H_1^{cT}(y,z)]\, dS \quad (4.40)$$

with

$$E_t^c(y,z) = (M_V^T)^{-1} E_t^m(y,z)$$
$$H_1^c(y,z) = (M_V^T)^{-1} H_1^c(y,z). \quad (4.41)$$

Introducing the substitutions (4.15) and (4.17) into the above equations shows that R, G, L, and C are defined in a unique way.

4.5.2 Characteristic impedance versus line-mode impedance

4.5.2.1 Characteristic impedance matrix In order to pursue the analogy between the high-frequency model defined by (4.25) and (4.30) and the quasi-TEM or TEM description still further, we now investigate whether it is possible to define an $N \times N$ characteristic impedance Z_c or characteristic admittance matrix Y_c with $Z_c = Y_c^{-1}$. These matrices express the relation between the circuit voltages and circuit currents propagating in the positive or negative x-direction:

$$V_c^{\pm}(x) = \pm Z_c I_c^{\pm}(x). \quad (4.42)$$

Z_c is independent of x and represents the input impedance matrix of the infinitely long multiconductor waveguide structure. Starting from (4.4) we have that

$$I_f^{\pm}(x) = \pm Z_f^{-1} V_f^{\pm}(x). \quad (4.43)$$

Replacing the field quantities $I_f^{\pm}(x)$ and $V_f^{\pm}(x)$ by circuit quantities, using

86   4 LOSSY MULTICONDUCTOR WAVEGUIDE STRUCTURES

(4.7) and (4.8), yields

$$M_I^{-1} I_c^\pm(x) = \pm Z_f^{-1} M_V^{-1} V_c^\pm(x) \qquad (4.44)$$

or

$$V_c^\pm(x) = \pm M_V Z_f M_I^{-1} I_c^\pm(x). \qquad (4.45)$$

Hence, the characteristic impedance matrix is given by

$$Z_c = M_V Z_f M_I^{-1}. \qquad (4.46)$$

Z_c in equation (4.46) is again defined in a unique way.

4.5.2.2 Line-mode impedance matrix The characteristic impedance matrix Z_c is sometimes confused with the line-mode impedance matrix Z_{lm} defined below. Element ij of this line-mode impedance matrix expresses the ratio between the circuit voltage $V_{c,ij}^\pm(x)$ on conductor i due to mode j and the circuit current $I_{c,ij}^\pm(x)$ on conductor i due to mode j:

$$Z_{lm,ij} = \pm \frac{V_{c,ij}^\pm(x)}{I_{c,ij}^\pm(x)} \qquad (4.47)$$

with $i, j = 1, 2, \ldots, N$. $V_{c,ij}^\pm(x)$ and $I_{c,ij}^\pm(x)$ are given by

$$\begin{aligned} V_{c,ij}^\pm(x) &= M_{V,ij} V_{f,j}^\pm(x) \\ I_{c,ij}^\pm(x) &= M_{I,ij} I_{f,j}^\pm(x). \end{aligned} \qquad (4.48)$$

Note that

$$\begin{aligned} V_{c,i}^\pm(x) &= \sum_{j=1}^N V_{c,ij}^\pm(x) \\ I_{c,i}^\pm(x) &= \sum_{j=1}^N I_{c,ij}^\pm(x). \end{aligned} \qquad (4.49)$$

Substitution of (4.48) in (4.47) yields

$$Z_{lm,ij} = \pm \frac{M_{V,ij}}{M_{I,ij}} \frac{V_{f,j}^\pm(x)}{I_{f,j}^\pm(x)} = \frac{M_{V,ij}}{M_{I,ij}} Z_{f,j}. \qquad (4.50)$$

Hence with (4.46) we get the following relation between the elements of Z_{lm} and Z_c:

$$Z_{lm,ij} = \sum_{k=1}^N Z_{c,ik} \frac{M_{I,kj}}{M_{I,ij}}. \qquad (4.51)$$

Note that it is not possible to write (4.51) in a closed form as a matrix relation. From (4.50) or (4.51) it can be seen that, as was the case for the circuit quantities, Z_{lm} is unique and does not depend on a specific choice of the arbitrary diagonal matrices δ_V and δ_I.

4.5.3 *Reciprocity of the full-wave circuit model*

We now come back to a problem which was first put forward in Section 1.2.8. One would expect that in the absence of non-reciprocal anisotropic media the circuit model for the considered multiconductor waveguides would be reciprocal. In other words one would expect that Z_c, z_c, and y_c, as well as R, G, L, and C, are symmetric matrices. However, in this section we will show that this is not necessarily the case. Only in the lossless case is this reciprocity guaranteed automatically.

We will concentrate on z_c; the calculations for y_c and Z_c are analogous and left to the reader. From (4.26) we have that z_c^{-1} is given by

$$z_c^{-1} = -M_I Z_f^{-1} j\beta^{-1} M_V^{-1}. \tag{4.52}$$

If we replace M_V with expression (4.14) we obtain

$$z_c^{-1} = -M_I Z_f^{-1} j\beta^{-1} (p^T)^{-1} M_I^{*T}. \tag{4.53}$$

Now we pick out an element ij with $i \neq j$:

$$(z_c^{-1})_{ij} = -\sum_{k=1}^{N} \sum_{l=1}^{N} M_{I,ik} Z_{f,k}^{-1} j\beta_k^{-1} (p^{-1})_{lk} M_{I,jl}^*. \tag{4.54}$$

We used the fact that Z_f and β are diagonal matrices. If z_c, and hence z_c^{-1}, has to be symmetric then element ij in (4.54) should be equal to element ji of z_c^{-1} given by

$$(z_c^{-1})_{ji} = -\sum_{k=1}^{N} \sum_{l=1}^{N} M_{I,jk} Z_{f,k}^{-1} j\beta_k^{-1} (p^{-1})_{lk} M_{I,il}^*. \tag{4.55}$$

Identifying the terms under the double summation on the right-hand side of (4.54) and (4.55) yields

$$M_{I,ik} M_{I,jl}^* = M_{I,jk} M_{I,il}^* \tag{4.56}$$

for $k, l = 1, 2, \ldots, N$. The reader might argue that only the double summations in (4.54) and (4.55) have to be identical and not each of the individual terms. However, we assumed the product $Z_{f,k} \beta_k$ to be different for all modes and to have no special relationship with each other. At the same time we made the assumption that the powers p_{lk} for fixed k values and different l values are independent. The only special relation that one could expect for the elements of p is between the elements p_{lk} and p_{kl}. However, this potential reciprocity relation does not come in here. We rewrite (4.56) as

$$\frac{M_{I,ik}}{M_{I,jk}} = \frac{M_{I,il}^*}{M_{I,jl}^*}. \tag{4.57}$$

In the special case that $k = l$ we have that

$$\frac{M_{I,ik}}{M_{I,jk}} = \frac{M^*_{I,ik}}{M^*_{I,jk}} \tag{4.58}$$

or that $M_{I,ik}/M_{I,jk}$ ($k = 1, 2, \ldots, N$) has to be a real quantity. Using the definition of M_I brings us to the condition that the quantity

$$\frac{\oint_{c_{c_i}} H^m_{t,k} \cdot \mathrm{d}\mathbf{l}}{\oint_{c_{c_l}} H^m_{t,k} \cdot \mathrm{d}\mathbf{l}} \tag{4.59}$$

has to be real. In a general lossy situation we know from equation (2.55) that $H^m_{t,k}(y,z)$ is a complex quantity with a non-constant phase over the cross-section of the waveguide. Hence, it is not very likely that (4.59) is real. Moreover, if we assume losses inside the conductors as well as for the surrounding media then it is not self-evident how to choose c_{c_i}. In that case c_{c_i} loses its preferential status. This means that, by taking another definition for c_{c_i}, we can change the numerator in (4.59) independently of the denominator. This proves that in a general situation it is not guaranteed that z_c is symmetric. For lossless structures we have, as is shown for the two-conductor case in the previous chapter, that p is a diagonal matrix due to the orthogonality of the modes. In that case condition (4.57) has only to be fulfilled when $k = l$. In other words, only (4.59) has to be real for $i, j, k = 1, 2, \ldots, N$. This is the case in lossless situations because $H^m_{t,k}(y,z)$ can always be chosen to be a real quantity.

Due to the approximations we made in constructing the power–current circuit model, we introduced non-reciprocity in this circuit model. The non-reciprocity finds its origin in the current condition and not in the power condition of the power–current formulation. This non-reciprocity is not a failure of the circuit model; it is just a consequence of the fact that we try to model a complex high-frequency electromagnetic structure with a simple set of coupled transmission lines. The same properties hold for a power–voltage circuit model, in which case the voltage condition will introduce non-reciprocity.

In many practical situations the non-diagonal elements of Z_c, z_c, y_c, R, G, L, and C are much smaller than the diagonal elements due to the weak electromagnetic coupling between the conductors. Moreover, the losses in practical waveguides are small, that is, the imaginary parts of ε and μ are small compared to the real parts. These two facts imply that in practical situations one does not notice much of the non-reciprocity.

Instead of demanding that M_I be defined by the current condition (4.7) and concluding that the resulting circuit model is not reciprocal, we could ask ourselves what condition has to be fulfilled for M_I in order to obtain a reciprocal circuit model. In other words we drop the condition (4.7) and keep only the power condition (4.13). We will now construct another relation between M_I and M_V which has to be fulfilled for a reciprocal circuit model. The matrices z_c (4.26) and y_c (4.26) are symmetric if the following two relations hold:

$$M_V \beta Z_f M_I^{-1} = (M_I^T)^{-1} Z_f \beta M_V^T$$
$$M_I \beta Z_f^{-1} M_V^{-1} = (M_V^T)^{-1} Z_f^{-1} \beta M_I^T. \tag{4.60}$$

Rewriting both conditions yields

$$A \beta Z_f = \beta Z_f A^T$$
$$A^T \beta Z_f^{-1} = \beta Z_f^{-1} A \tag{4.61}$$

with the $N \times N$ matrix A with elements a_{ik} defined as

$$A = M_I^T M_V. \tag{4.62}$$

If we write the conditions (4.61) element by element and use the fact that β and Z_f are diagonal matrices we obtain

$$a_{ik} \beta_k Z_{f,k} = \beta_i Z_{f,i} a_{ki}$$
$$\frac{a_{ki} \beta_k}{Z_{f,k}} = \frac{\beta_i a_{ik}}{Z_{f,i}} \tag{4.63}$$

with $i, k = 1, 2, \ldots, N$. If $k \neq i$ and if $\beta_k \neq \beta_i$, then can only be fulfilled if $a_{ki} = 0$. In other words, the circuit model will be reciprocal if

$$M_I^T M_V = D \tag{4.64}$$

with D a diagonal $N \times N$ matrix. Conditions (4.13) and (4.64) do not determine the circuit model fully as conditions (4.7) and (4.13) did. This is due to the fact that the elements on the diagonal of D are arbitrary. It is left to the reader as an exercise to show that the condition (4.64) is equivalent to condition (4.57).

Observe that (4.64), that is, the reciprocity condition, is automatically guaranteed by (4.13), that is, the power conservation principle, if no losses are present. In that case p is diagonal due to the orthogonality of the modes and M_I can be chosen to be real.

4.6 Case study

To illustrate this chapter we analyse the asymmetric two-conductor lossy microstrip structure shown in Fig. 4.2. The conductors are perfectly conducting and have a thickness $t = 0.05$ mm. The width of the left conductor

90   4 LOSSY MULTICONDUCTOR WAVEGUIDE STRUCTURES

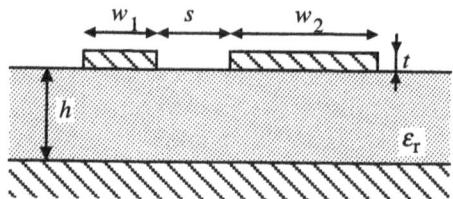

Fig. 4.2. Geometry of an asymmetric two-conductor microstrip structure on top of a lossy substrate.

w_1 is 1 mm and the width of the right conductor w_2 is 2 mm. The spacing s between the conductors is 1 mm. The substrate has a thickness $h = 0.635$ mm. The real part of the relative permittivity of the substrate is given by $\text{Re}(\varepsilon_r) = 9.8$ and the loss tangent is given by $\text{tg } \delta = 0.05$. Hence, the complex permittivity is equal to $\varepsilon_r = 9.8(1 - 0.05j)$.

We analysed the structure with the full-wave techniques described in Chapter 9. The normalized propagation constants of the two fundamental modes (the so-called c and π modes) as a function of frequency are shown in Fig. 4.3. Figure 4.4 shows the real and imaginary parts of the elements of the line-mode impedance matrix:

$$Z_{lm} = \begin{pmatrix} Z^{(c)}_{lm,1} & Z^{(\pi)}_{lm,1} \\ Z^{(c)}_{lm,2} & Z^{(\pi)}_{lm,2} \end{pmatrix}. \tag{4.65}$$

The characteristic impedance matrix Z_c is shown in Fig. 4.5. Note the considerable difference between the values of the line-mode impedances and

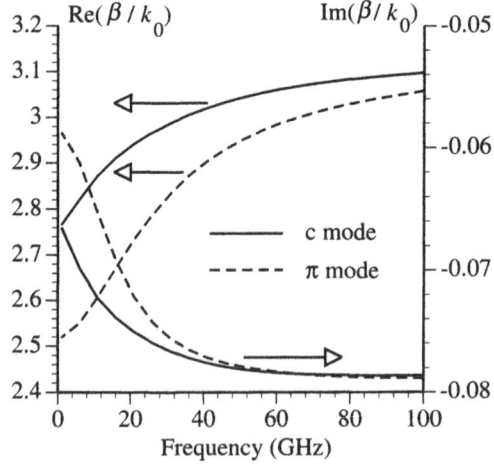

Fig. 4.3. Normalized complex propagation constants of the π and c mode.

4.6 CASE STUDY

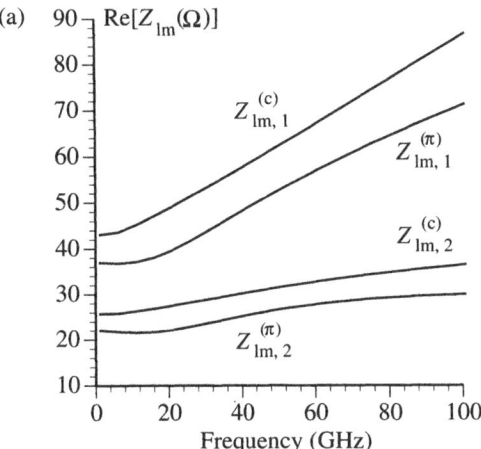

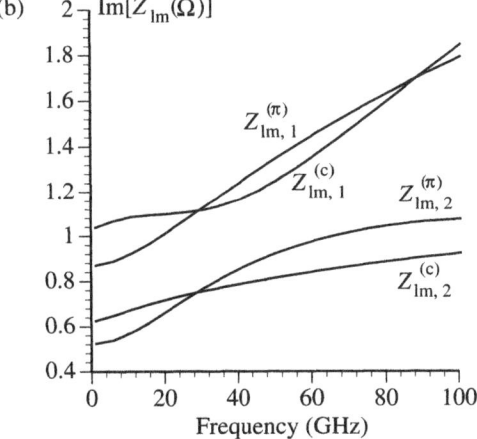

Fig. 4.4. Real and imaginary parts of the elements of the line-mode impedance matrix $\mathbf{Z}_{lm}(\omega)$: (a) real parts; (b) imaginary parts.

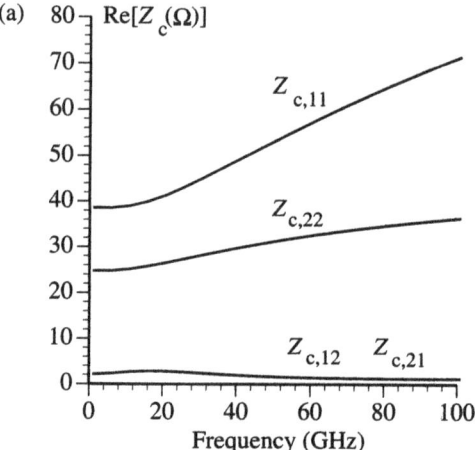

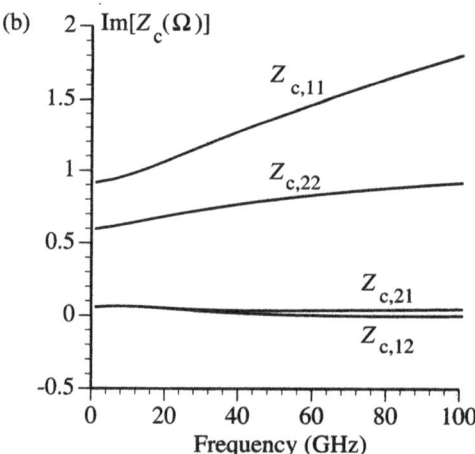

Fig. 4.5. Real and imaginary parts of the elements of the characteristic impedance matrix $Z_c(\omega)$: (a) real parts; (b) imaginary parts.

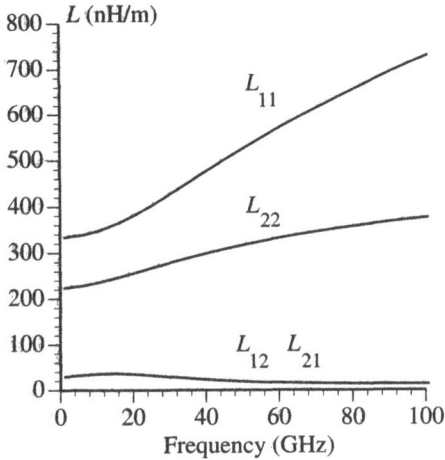

Fig. 4.6. Elements of the inductance matrix $L(\omega)$.

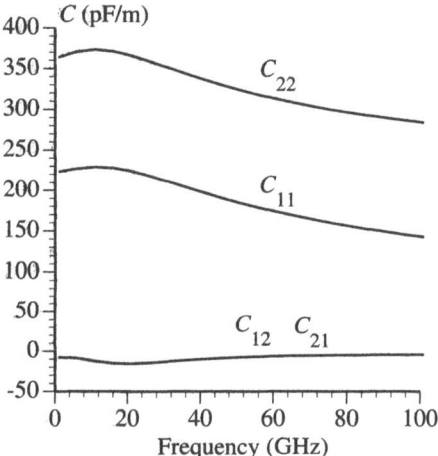

Fig. 4.7. Elements of the capacitance matrix $C(\omega)$.

the characteristic impedances. At the low-frequency end, for example, Fig. 4.5 shows that the coupling between the lines is small and that line 1 has a characteristic impedance of about $(39 + 0.9j)\Omega$ and line 2 an impedance of $(25 + 0.6j)\Omega$. The line-mode impedances would lead us to a quite different but erroneous conclusion. Finally Figs 4.6, 4.7, and 4.8 show the frequency-dependent $L(\omega)$, $C(\omega)$, and $G(\omega)$ matrices. The elements of the $R(\omega)$ matrix are negligible because there are no losses inside the conductors. Note that

94　4　LOSSY MULTICONDUCTOR WAVEGUIDE STRUCTURES

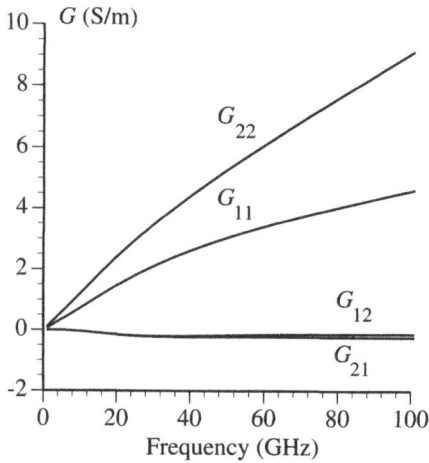

Fig. 4.8. Elements of the conductance matrix $G(\omega)$.

even with the considerable losses in the substrate the non-reciprocity in G and $\mathrm{Im}[Z_c]$ is hardly noticeable even at higher frequencies.

References

[1] Faché, N. and De Zutter, D. (1990). New high-frequency circuit model for coupled lossless and lossy waveguide structures. *IEEE Transactions on Microwave Theory and Techniques*, **MTT-38**, 3, 252–259.
[2] Olyslager, F., Dhaene, T., and De Zutter, D. (1992). Time and frequency domain study of the propagation in lossy multilayered waveguide and antenna structures based on a rigorous full-wave analysis. In *Proceedings of the 1992 IEEE-APS International Symposium*, pp. 1504–1507, Chicago.
[3] Dhaene, T. and De Zutter, D. (1992). CAD-oriented, general circuit description of uniform coupled lossy dispersive waveguide structures. *IEEE Transactions on Microwave Theory and Techniques*, **MTT-40**, 7, 1545–1554.
[4] Dhaene, T., Olyslager, F., De Zutter, D., and Lagasse, P. (1992). Transmission line modelling of coupled lossy open dispersive hybrid waveguide structures in isotropic and anisotropic media. In *Proceedings of the 1992 URSI International Symposium on Signals, Systems and Electronics*, pp. 427–429, Paris.
[5] Lindell, I. V. (1981). On the quasi-TEM modes in inhomogeneous multiconductor transmission lines. *IEEE Transactions on Microwave Theory and Techniques*, **MTT-29**, 8, 812–817.
[6] Lindell, I. V. and Gu, Q. (1987). Theory of time-domain quasi-TEM modes in inhomogeneous multiconductor transmission lines. *IEEE Transactions on Microwave Theory and Techniques*, **MTT-35**, 10, 893–897.
[7] Brews, J. R. (1986). Transmission line models for lossy waveguide interconnections in VLSI. *IEEE Transactions on Electronic Devices*, **ED-33**, 9, 1356–1365.

5
GENERAL INTRODUCTION TO THE FULL-WAVE ANALYSIS OF MULTICONDUCTOR LINES IN A PLANAR STRATIFIED MEDIUM

5.1 Full-wave analysis of two- and three-dimensional waveguide structures embedded in a planar stratified medium

In the first part of this book we examined in detail how a circuit model can be derived for a set of multiconductor transmission lines. The discussion was restricted to the propagating fundamental modes. The equivalent circuit model turned out to be a set of coupled transmission lines. The relation between the typical circuit quantities, that is, voltages, currents, (coupling) impedances, and signal velocities, and the original field quantities, that is, the modal fields and associated modal propagation constants, was determined.

For microstrip and stripline types of multiconductor structures the circuit models are based on the modal propagation constants, on the power propagated by the actual structure, and on the total longitudinal current flowing along each conductor. The modal propagation constants also become the propagation constants of the modes in the circuit model. The power equivalence principle ensures that the complex power propagated by the waveguide structure is the same as the one propagated by the circuit model. Finally, we identified the total longitudinal current flowing along each conductor with the circuit current. In this way we arrived at the power–current definition of the impedances.

For a slotline type of waveguide, it is certainly more advantageous to identify the potential difference over the slot with the circuit voltage. In that case the impedances are defined by a power–voltage approach.

From a technological point of view, extremely important types of multiconductor lines are those where the conductors are embedded in a multilayered or, more precisely, a planar stratified medium. This is a subclass of the more general structures dealt with in the first part of this book.

In the second part of the book, three types of the above subclass will be investigated. The final purpose of this study is to determine the necessary quantities to derive a circuit model for these structures. The circuit examples given in the first part of the book were in fact based on the full-wave analyses that will be presented in Chapters 6–9.

In this introduction the work presented in Chapter 6–9 is placed within its proper context with respect to the state-of-the-art in the full-wave analysis of two- and three-dimensional multiconductor waveguide structures. Some of the references given here will also be found in the subsequent chapters. This is a consequence of our choice of making each chapter into a largely self-contained unit.

We also intend to indicate very briefly the technological relevance of each of the structures examined. Furthermore, the general outline of the theory developed in each chapter is put forward with special emphasis on the different lines of thought that oriented our research over the last five years.

It is certainly not our purpose to present a comprehensive review of the various waveguide structures that are used in present day technology and of the very many methods that exist in order to model the wave propagation along these structures. We refer the reader to the extremely abundant literature on these topics and in particular to [1] for an up-to-date overview of both the numerical methods used for, and the most important papers dealing with, the modelling of two-dimensional multiconductor transmission lines and three-dimensional discontinuity problems. We would also like to mention one additional comprehensive source of published research results provided by the best specialists. We refer to the PIER (Progress in Electromagnetic Research) series. Five volumes have been published in this series so far [2].

The expertise of the present authors, which forms the basis for this book, is mainly in the field of the integral equation formulation of electromagnetic field problems and their solution based on the method of moments either in the space domain or in the spectral domain. As already emphasized in [1] no single numerical method is or will turn out to be superior over other ones. Much depends on the particularities of the problem at hand. However, the continuous increase in available computing power paves the way for very general methods such as finite differences, finite elements, or the transmission line method that require no, or only a small amount of, precalculation or analytical investigations. Generally speaking, the advantage of an integral equation approach resides in the fact that the fields in a three-dimensional structure can be derived from unknown field quantities on certain boundaries. In the case of two-dimensional waveguide problems these boundaries reduce to the circumferences of the conductors. This of course substantially reduces the number of unknowns as compared to the finite element method or finite difference method where the whole two-dimensional cross-section of the multiconductor structure must be discretized. Of course, a price has to be paid for this advantage: only piecewise homogeneous media can be handled, a suitable Green's function must be found, and rather extensive analytical preprocessing is required (at least in those cases presented below). Before concluding this rather general discussion on the numerical approach adopted

5.2 INFINITELY THIN STRIP CONDUCTOR LINES

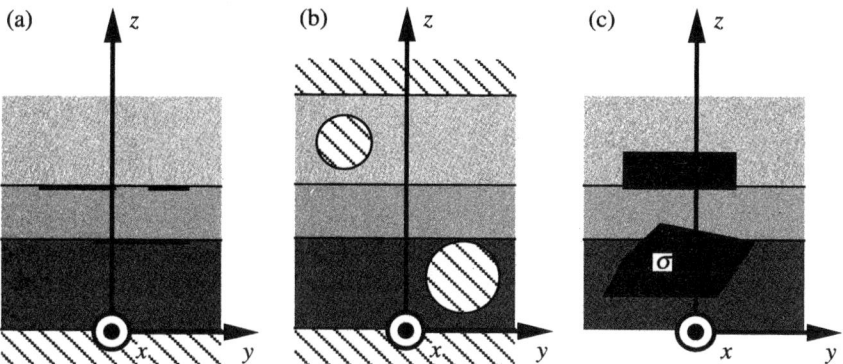

Fig. 5.1. The three types of multiconductor lines treated in this book: (a) infinitely thin perfectly conducting strips; (b) perfectly conducting wire transmission lines; (c) arbitrarily shaped polygonal conductors with finite conductivity.

in this second part, we draw the attention of the reader to the fact that the emergence of software packages for the symbolic treatment of mathematics constitutes a valuable help in this analytical preprocessing [3].

The three types of multiconductor lines treated in this book are:

(1) infinitely thin perfectly conducting strips;
(2) perfectly conducting wire transmission lines; and
(3) arbitrarily shaped polygonal conductors with finite conductivity.

As we consider only the two-dimensional waveguide problem, the cross-section of these three types of multiconductor lines is invariant along the propagation direction. A general feature of each of these transmission line types resides in the fact that they are embedded in a multilayered medium consisting of an arbitrary number of homogeneous and isotropic layers as shown in Fig. 5.1. If this multilayered medium is placed on top of a ground plane, we talk of a microstrip-like configuration (Fig. 5.1(a)). The common microstrip consisting of a ground-plane–substrate–air configuration is certainly the most well-known representative of this family. In case the multilayered medium is sandwiched between ground planes, we talk of a stripline-like configuration (Fig. 5.1(b)). A third type of structure is the open one without ground planes (Fig. 5.1(c)).

Let us now take a closer look at each class of multiconductor lines that we will treat.

5.2 Infinitely thin strip conductor lines

The infinitely thin strip conductors constitute a good model for the interconnections used in microwave and millimetre-wave integrated circuits

(MICs) or monolithic microwave integrated circuits (MMICs). Most MICs use low-loss dielectrics as a substrate (of which Duroid is a good example). MMICs on the other hand use semi-insulating substrates as GaAs because these substrates are suited for the fabrication of active circuits. We refer the reader to [4] for a comprehensive technological overview. The increasing importance of optical interconnection techniques has led to the fabrication of OEICs (Opto-Electronic Integrated Circuits) [5]. In a typical communication system the optical signal has to be translated into a digital electrical signal. Here again microstrip or stripline types of interconnections play a key role.

It is certainly not economic, and even impossible, to tune these integrated circuits. This calls for reliable and accurate CAD tools to predict the behaviour of such circuits. Existing (microwave) design packages [6], [7] are based on the availability of the circuit parameters (either S-parameters, Y-parameters, or some other equivalent representation) of a large number of basic building blocks such as interconnecting lines, discontinuities, and active circuits. Although the available computer power has increased dramatically in recent years, accurate electromagnetic characterization is often too slow to be incorporated directly into an optimizing CAD tool. However, the importance of these numerical methods resides in the fact that they serve as a basis for curve fitting or for validating empirical formulas. A number of commercially available planar structure simulators or three-dimensional waveguide problem solvers also serve as direct sources for the S-parameter calculation of more complex building blocks.

In practice, the infinitely thin character of the conductors discussed in Chapter 7 is of course an approximation, together with their perfectly conducting character. The latter property can be weakened by introducing a surface impedance which determines the ratio between the tangential electric and magnetic fields on the strip [8].

By assuming that all conductors embedded in a multilayered medium are infinitely thin, the full-wave analysis of the problem can be substantially simplified. Since its introduction in 1973 by Itoh and Mittra, the spectral domain method became a powerful tool for the solution of thin strip problems. We again refer the reader to [1] for a short but very clear introduction to this method. The spectral domain method explicitly requires infinitesimal thickness for the strip conductors and finite conductivity is not easily handled. In essence, the method amounts to transforming the integral equation that governs the problem to the spectral domain. In order to do so a Fourier transform is taken along the y-axis of Fig. 5.1. By applying Galerkin's method [9] a homogeneous system of equations is obtained from which the modal propagation constants and the modal surface current distribution can be derived.

The approach presented in Chapter 7 and in [10]–[13] differs from the spectral domain approach. The full-wave solution is formulated in the original space domain rather than in the spectral domain. It starts from the calculation of a Green's dyadic (that is, a matrix of Green's functions) in the spectral domain and proceeds by explicitly inverse-Fourier transforming the dyadic. Both the transverse and longitudinal currents are discretized using the method of moments in such a way that the edge behaviour is satisfied. This edge behaviour of the fields and currents, first formulated by Meixner [14], states that the longitudinal current on each strip becomes infinite near the strip edge as the reciprocal of the square root of the distance to the edge. The transversal current on the other hand becomes zero in a way which is proportional to the square root of that same distance.

The boundary condition (zero tangential electric field) on each strip is not imposed in a mean sense as in Galerkin's method, but at a number of points equally spaced along each strip, that is, we solve the integral equation by applying the point matching or collocation method. The proposed method is very powerful in handling coupled line problems, especially those where the coupling is rather strong and where great flexibility in the surface current modelling is essential in order to correctly predict the interaction between neighbouring strips. It is also very well suited to examining the influence of the frequency on the current.

We certainly do not pretend that the solution of the integral equation in the space domain is a better alternative than the spectral domain technique but we are convinced that the space domain approach is certainly a worthwhile alternative.

A distinguishing feature of the modelling approach presented in the case of thin conductors is that the layered medium is accounted for by selecting a Green's dyadic as the kernel of the integral equation [15]. This dyadic ensures that Maxwell's equations, together with the associated boundary and/or radiation conditions, are satisfied throughout the layered medium. This implies that radiation phenomena (both surface and space waves) are included. For thin conductors, the electric field due to an elementary surface current located in a plane parallel to the (x, y)-plane is the required Green's dyadic.

5.3 Spectral domain description of a planar stratified medium

For wire transmission lines and for arbitrary polygonal waveguide structures, we will need the fields scattered by the layered medium for a given incident field defined in one of the layers. In all cases a special technique must be developed to handle the layered medium. By introducing a Fourier transform of all field components with respect to the y-axis, that is, by passing to the spectral domain, Maxwell's equations for each propagating mode can be solved analytically in the spectral domain. Because this spectral treatment of the layered medium is so crucial, Chapter 6 has been devoted to it. We have chosen to present this before the chapters that actually deal with the multiconductor waveguides.

The problem of electromagnetic propagation in stratified media, both isotropic and anisotropic, has been extensively studied in the past. A review of relevant papers can be found in [16]–[18]. For three-dimensional problems the presence of a layered medium in most cases leads to the calculation of Sommerfeld integrals [1]. In the two-dimensional case that is of interest here, the space domain results can be derived directly from the spectral domain results by inverse Fourier transformation. The integrands of these Fourier integrals exhibit the same difficulties as their counterparts in the three-dimensional case, that is, the presence of branch cuts and poles on or close to the integration path. Moreover, special care has to be taken to correctly handle large absolute values of the spectral variable corresponding to field values in the immediate neighbourhood of the sources.

The purpose of Chapter 6 is to provide the reader with the necessary insight and with the mathematics to confidently handle the layered medium problem, at least in the two-dimensional case. Transmission line concepts are introduced to provide an easy to understand interpretation of the wave phenomena. The numerical procedures are cast in such a form that rounding errors due to the presence of losses and the associated exponential decay of the waves are avoided. A similar technique, but for the three-dimensional case and with special application to microstrip antennas, was presented in [18].

Finally, special attention is also devoted to power calculations in the layered medium. Indeed, after solving the integral equation governing the modal field problem, we still have to determine the power propagated by each mode and in the case of a multimode structure we also need the power propagated by the electric field of one mode and the magnetic field of another mode. In the following we will refer to the power defined by fields of different modes with the term cross-power. The convolution theorem of Fourier analysis leads to an elegant way of directly determining these power expressions in the spectral domain. Chapter 6 concludes with the relevant expressions for these power terms.

5.4 Discrete wire lines

A second type of multiconductor transmission line described in this book is the wire transmission line. Many readers will be much less familiar with this type of transmission line as compared to microstrip and stripline configurations. Two important applications of discrete wire technology, the Multiwire®

5.4 DISCRETE WIRE LINES

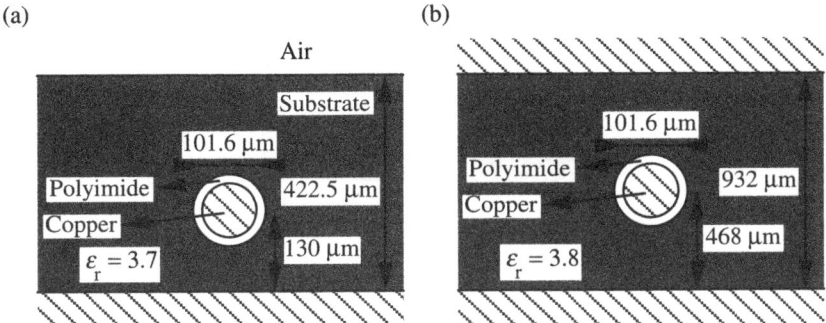

Fig. 5.2. Multiwire controlled-impedance transmission lines: (a) microstrip type (50 Ω); (b) stripline type (75 Ω).

and Microwire® circuit boards, have created a new technology for printed circuit boards [4], [19], [20]. The main difference between these and the classical printed circuit boards resides in the fact that no etching techniques are used to define interconnection patterns. The discrete wire technology uses isolated copper wires that are deposited by a specially constructed wiring machine. These wires are embedded inside a layered medium (see Fig. 5.2). The diameter of the wire can take several values but 101.6 μm (4 mil) (without isolation) and 63.5 μm (2.5 mil) are the most frequently used. Both the (buried) microstrip (Fig. 5.2(a)) and stripline type of configuration (Fig. 5.2(b)) can be fabricated. By carefully controlling the distance between the centre of the wire and the ground plane(s) an impedance-controlled transmission line is obtained providing the necessary electrical characteristics to transport high-speed digital signals. As in classical printed circuit boards, wire transmission lines of the microstrip and stripline type can be combined into a multilayered board. Additional technological details can be found in [4]. One of the main features that we want to mention here is that an impedance control of up to 5 per cent of the nominal value (typically 50 Ω or 75 Ω) is possible whereas classical multilayered boards only allow an impedance control of about 10 per cent. Rapid technological changes can of course invalidate this comparison, certainly with the increasing demands coming from the emergence of broadband digital networks.

In view of the increasing bit-rates (about 1 Gbit/s) and the use of discrete wire technology in microwave applications, it becomes interesting to investigate the full-wave properties of that technology. The low-frequency or quasi-TEM properties had already been investigated by Shibata and Terakado [21] and in a number of papers dealing with the general problem of the quasi-TEM analysis of multiconductor transmission lines, including two papers by some of the present authors [22]–[24]. In the situation sketched on Fig. 5.3, where the wire is located above a substrate on top of a ground

5 FULL-WAVE ANALYSIS OF MULTICONDUCTOR LINES

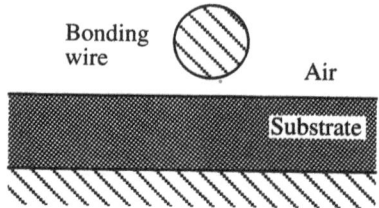

Fig. 5.3. Modelling of a bonding wire above a substrate.

plane, the wire can also serve as a model for a bonding wire in the packaging of high-speed integrated circuits [25].

The analysis presented in Chapter 8 is based on the first published full-wave analyses of single and coupled discrete wire structures [26], [27]. As for the thin strip conductors the perfectly conducting wires are embedded in a multilayered medium. The modal field distributions and modal propagation constants are again found by solving an integral equation for the surface currents on each wire. In the actual development of the theory the two scalar components of the surface current (longitudinal x component and azimuthal ϕ component, see Fig. 5.4) are replaced by an equivalent set of basis functions, that is, the longitudinal component of the magnetic field and the normal derivative of the longitudinal component of the electric field. The boundary conditions enforced by this integral equation on the surface of each wire require the longitudinal component of the electric field to be zero and the longitudinal component of the magnetic field to be the opposite of the azimuthal current. These are of course the necessary conditions for a perfect conductor.

In contrast to the approach adopted for thin strip conductors, the Green's dyadic for the layered medium is not used as such! We now start from a

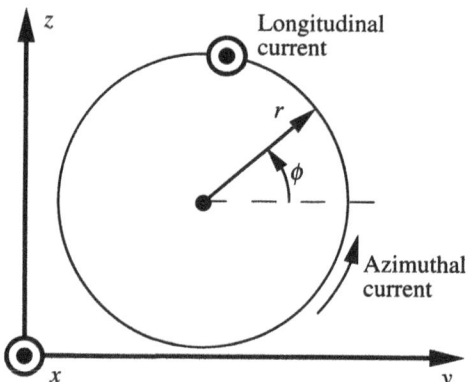

Fig. 5.4. Components of the surface current on a perfectly conducting wire.

Fourier series representation of the unknowns on the surface of each wire. The fields generated everywhere inside the layered medium are found by considering these fields as the superposition of incoming field contributions and scattered field contributions. The incoming fields are associated with each individual wire and are defined as those fields that would be generated by the surface currents on that wire if the layer in which the wire resides were replaced by a homogeneous space of infinite extent with the same material parameters as the original layer. The scattered fields represent the 'reaction' due to the presence of the stratified medium. For the scattered fields an explicit transformation to the spectral domain and from the spectral domain back to the original space domain is invoked. As for thin strip conductors, this appeal to the spectral domain is inevitable in the presence of layered media. The boundary conditions on each wire are enforced on a per-Fourier-term basis.

From the above it can be concluded that we in fact construct the response of the layered medium for a specific excitation, that is, for a current on a cylindrical surface and with an $e^{jn\phi}$ dependence ($n = 0, \pm 1, \pm 2, \ldots$). This special form of the current is explicitly exploited to maximize the benefit from the analytical precomputations. In the actual numerical implementation of the method, the Fourier series has to be truncated to a finite number of terms. Technically speaking this means that the integral equation (or the coupled set of integral equations in the case of coupled wires) is solved using a finite set of Fourier terms as basis and test functions. This is the well-known Galerkin method.

In comparison to the situation for thin strip conductors, the power calculations become somewhat more complicated and time-consuming. Resorting to the convolution technique described in Chapter 6 is possible only in those layers where no wire is present. In the other layers the Poynting vector must be explicitly integrated in the space domain.

5.5 Multiconductor lines formed by polygonal-shaped conductors with finite conductivity

The third and last type of multiconductor transmission line structure analysed in this book consists of conductors with finite conductivity and with polygonal cross-section (see Fig. 5.5). It is of course only natural to consider the extension of very thin conductors to structures of finite thickness and finite conductivity. This enables one to assess the influence of these parameters on the modal characteristics. Technologically important issues such as underetching, which influences the actual shape of the conductor, and conductor roughness, which is one of the factors that contribute to losses, can be investigated. Another very important issue to investigate is the way

104   5 FULL-WAVE ANALYSIS OF MULTICONDUCTOR LINES

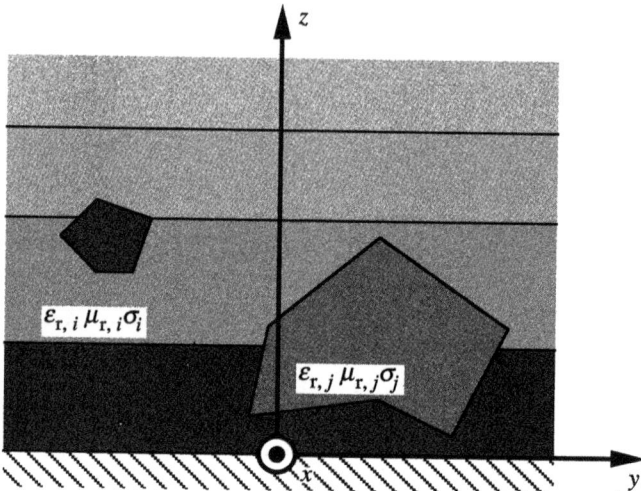

Fig. 5.5. Multiconductor line formed by polygonal-shaped conductors with finite conductivity.

in which the fields penetrate the conductor as a function of frequency. If the thickness of the conductor is sufficiently high compared to the wavelength, it is assumed that the skin effect theory can be safely applied. However, if the wavelength becomes too large or conversely if the conductor thickness becomes too small, the fields will penetrate into the conductor in a way which can no longer be accounted for by that theory.

Multiconductor transmission lines consisting of polygonal-shaped conductors embedded in a multilayered medium have already been studied extensively in the quasi-TEM limit [22], [23]. A recent contribution by the present authors takes the singular behaviour of the charges in the neighbourhood of conductor corners into acount [24]. The capacitance and inductance matrices associated with these multiconductor lines, and hence the modal characteristics, can be calculated in a very accurate and efficient way. By virtue of the approximations that are inherent to the quasi-TEM theory, conductors must be perfectly conducting. However, finite conductivity can be taken into account by a suitable perturbation technique based on the skin effect. Lossy layers can also be included provided the loss tangent of the dielectrics is sufficiently small.

The full-wave analysis of multiconductor lines is much more complicated and only a very limited number of authors have tackled the problem. The analysis of the behaviour of the propagation constants of the fundamental and higher-order modes of perfectly conducting polygonal-shaped waveguides was presented by Michalski and Zheng [28]. Chapter 9 presents the first full-wave analysis of finite conductivity lines embedded in a stratified medium

[29]–[31]. Both single and coupled lines are discussed. The effect of conductivity is taken into account in an exact way. No perturbation techniques are required. Moreover, the limiting case of perfectly conducting lines is included as well. We not only restrict ourselves to the determination of the complex propagation constants of the modes but also present results for the (coupling) impedances by calculating the (cross-) power propagated by the different modes. This allows us to realize the goal put forward at the beginning of this introduction, that is, to determine the necessary data in order to construct a full-wave circuit model based on the theory presented in the first part of the book. The proposed analysis remains valid for conductors that are partially embedded in one layer and partially in another layer.

We conclude this introduction by enumerating the salient features of the full-wave analysis for conducting transmission lines.

The integral equation is constructed by taking the tangential electric and magnetic fields at the surface of each conductor as the unknowns of the problem. The problem could be formulated in a totally equivalent way by introducing unknown electric and magnetic surface currents on these boundary surfaces. A Fourier series expansion of these tangential fields, as for the wires, is of course not appropriate. In this case, the application of the method of moments is based on a representation of the longitudinal tangential field components (E_x and H_x) as piecewise linear functions along the circumference of each conductor, while the other tangential components situated in the (y, z) plane (see Fig. 5.6) are represented by piecewise constant functions. In the following we will refer to the latter components as the transversal tangential components.

As for the wires, the still unknown fields on the surface of the conductors are used to determine incoming and scattered fields, not only everywhere inside the layered medium but also inside the conductors. The latter can be directly found in the space domain using the scalar Green's function (and its derivative) of the homogeneous conductor material. The fields in the stratified medium, in particular the scattered ones, are again obtained by switching to the spectral domain. Finally, the integral equation that governs the physics of the problem is found by enforcing the continuity of the tangential electric and magnetic fields at the surface of each conductor when approaching that surface first from inside the layered medium and then from inside the conductor.

The integral equation is not enforced at every point but only in a mean sense over each of the elementary intervals in which the circumference of each conductor has been divided. This is done by applying Galerkin's procedure, that is, by introducing a set of weighting or test functions. For the longitudinal field components the test functions are chosen to be piecewise constant over the same intervals in which the basis functions were

106   5 FULL-WAVE ANALYSIS OF MULTICONDUCTOR LINES

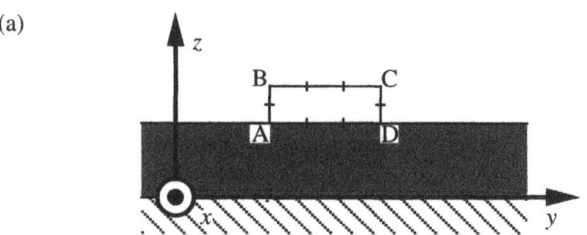

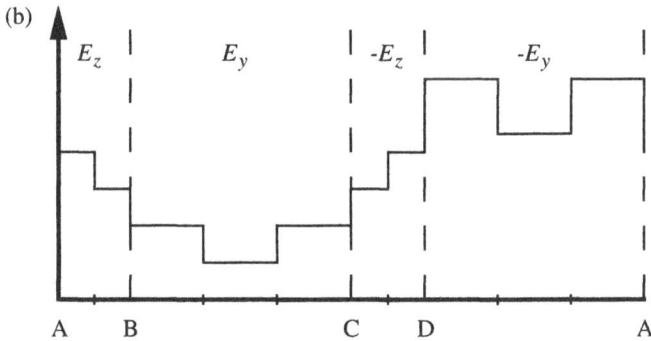

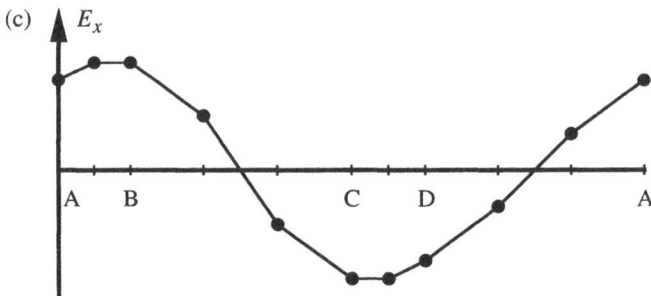

Fig. 5.6. Representation of the tangential electric field on the conductor surface: (a) conductor example; (b) piecewise linear representation of the longitudinal electric field; (c) piecewise constant representation of the transversal tangential electric field.

defined. For the transversal tangential field components the test functions are chosen to be piecewise linear. This choice implies that the basis functions for the longitudinal components become the test functions for the transversal components and vice versa.

In the analytical precomputations the integration over each basis function and the integration over each test function are performed analytically. The basis function integration extends over each elementary source interval where

5.6 FEATURES OF THE INTEGRAL EQUATION APPROACH

source means the tangential electric or magnetic field on the surface of each conductor. The test function integration extends over each elementary observation interval where observing amounts to taking the mean value by multiplication with the prescribed weighting function of a tangential field component, again on the surface of each conductor. This procedure only turns out to be possible in the spectral domain. The only integrals that have to be calculated numerically are inverse Fourier transformations. This implies that our method is in fact a spectral domain method. In contradistinction to the original spectral domain method, however, we are not restricted to infinitely thin conductors.

5.6 Additional features of the integral equation approach

We finally want to draw the attention of the reader to some other features of the approaches described above.

5.6.1 *Modelling of the singular behaviour of the current near conductor edges*

In the case of perfectly conducting multiconductor lines, the tangential electric field vanishes and the tangential magnetic field becomes the sole unknown. By virtue of the boundary conditions at a perfect conductor, the longitudinal magnetic field corresponds to the transversal surface current while the transversal magnetic field corresponds to the longitudinal current. Considering the thin strip conductors of Chapter 7 as a special case of the more general conductors handled in Chapter 9, this implies that the longitudinal current on such thin conductors is represented by a piecewise constant function, whereas the transversal current is represented by a piecewise linear function. This piecewise linear function would obviously have to be zero at the edges of the strip. In Chapter 7, however, we opted for a more sophisticated modelling by taking into account the singular behaviour of the current near the edges. This leads to substantial analytical complications that are deliberately avoided in the already intricate general conductor case.

In connection with the explicit modelling of the singular behaviour of currents near conductor edges, two additional considerations should be taken into account. In the implementation of the integral equation for the general conductor case, a conductor with a smooth boundary is approximated by a polygonal-shaped conductor. The special behaviour of the current near the edges of that polygon is certainly not present in the original conductor and could lead to inaccurate results.

A second remark can best be formulated by looking at the example given in Fig. 5.7, which shows a perfectly conducting thick strip with a width

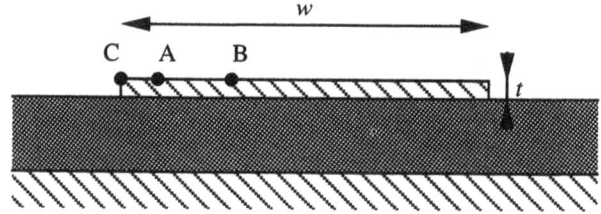

Fig. 5.7. Current singularity near a (perfect) conductor edge.

w much bigger than its thickness t. Close to the corner (in the part CA where CA is of the order of t) and according to Meixner [14] we expect the longitudinal (that is, z-directed) current to behave in a $\tau^{-\frac{2}{3}}$ way where τ is the distance to the corner C. This is due to the fact that on the geometrical scale dictated by the thickness t the influence of the corner C is most certainly that of a 90° angle. However, somewhat further away from C, in the interval AB, for example, the geometry of the problem is determined more by the fact that t is very small with respect to w and that the conductor very much resembles an (infinitely) thin conductor with its typical $\tau^{-\frac{1}{2}}$ singularity for the longitudinal current near the edge. The above example sufficiently illustrates the problem of the correct current modelling near conductor edges. For that reason we prefer to adopt the more cautious procedure of representing the longitudinal current in a piecewise constant way and to let Maxwell's equations decide how the final current profile will look.

5.6.2 Extended spectral domain approach

We systematically use the spectral domain to take the presence of the stratified medium into account. To this end a Fourier transformation and its inverse along the y-axis in Fig. 5.5 are introduced. However, a special situation occurs when two conductors are embedded in the same layer. In that case the contribution of the incoming fields from one conductor to the total fields on the other conductor is obtained by introducing a Fourier transformation along a direction not coinciding with the y-axis but roughly speaking with a direction orthogonal to the line connecting the centres of both conductors. A more rigorous explanation of the technique can be found in Chapter 9, but it is worthwhile to notice that this modified spectral domain technique allows us to circumvent a lot of numerical difficulties. The same technique was used for the coupling between wires embedded in the same layer.

We hope that by now the reader has obtained a clear picture of the basic principles and choices that were adopted in our full-wave analyses. Starting out with an explicit space domain alternative to the more familiar spectral domain analysis for thin strips, our technique evolved towards an extended spectral domain approach in the case of conductors with finite conductivity. We now lead the reader to a separate account of each of the full-wave problems discussed above, preceded by a chapter on spectral field calculations in a planar stratified medium.

References

[1] Itoh, T. (ed.) (1989). *Numerical Techniques for Microwave and Millimeter-Wave Passive Structures.* Wiley, New York.

[2] Jin Au Kong (ed.) (1989–91). *Progress in Electromagnetics Research.* Volumes 1–5. Elsevier, New York.

[3] Silvester, P. (1991). Symbolic computation as a basis for numerical methods. In *International Conference on Computation in Electromagnetics.* IEEE Conference Publication 350, pp. 1–5, London.

[4] Coombs, C. F. (ed.) (1988). *Printed Circuits Handbook* (3rd edn.). McGraw-Hill, New York.

[5] Schlafer, J. and Lauer, R. B. (1990). Microwave packaging of optoelectronic components. *IEEE Transactions on Microwave Theory and Techniques*, **MTT-38**, 5, 518–523.

[6] Pitzalis, O. (1989). Microwave to mm-wave CAE: concept to production. *Microwave Journal.* 1989 State of the Art Reference, 15–47.

[7] IEEE Spectrum, *Focus Report on Software*, November 1991.

[8] Peng, S. T., Tzuang, C-K. C., and Chen, C. D. (1990) Full-wave analysis of lossy transmission lines incorporating the metal modes. In *1990 IEEE MTT-S International Symposium Digest*, Vol. I, pp. 171–174, Dallas, Texas.

[9] Harrington, R. F. (1968). *Field Computations by Moment Methods.* Macmillan, New York.

[10] Faché, N. and De Zutter, D. (1988). Rigorous full-wave space-domain solution for dispersive microstrip lines. *IEEE Transactions on Microwave Theory and Techniques*, **MTT-36**, 4, 731–737.

[11] Faché, N. and De Zutter, D. (1988). Full-wave space domain solution for microstrip lines. In *Proceedings of the 1988 Antennas and Propagation Symposium and URSI Radio Science Meeting*, pp. 1026–1029, Syracuse, USA.

[12] Faché, N. and De Zutter, D. (1989). Circuit parameters for single and coupled microstrip lines by a rigorous full-wave space-domain analysis. *IEEE Transactions on Microwave Theory and Techniques*, (special issue on quasi-planar millimeter-wave components and subsystems), **MTT-37**, 2, 421–425.

[13] Faché, N., Van Hese, J., and De Zutter, D. (1989). Space domain Green's dyadic for non-coplanar microstrip or striplines in multilayered media. In *Proceedings of the 1989 URSI International Symposium on Electromagnetic Theory*, pp. 378–380, Stockholm.

[14] Meixner, J. (1972). The behaviour of electromagnetic fields at edges. *IEEE Transactions on Antennas and Propagation*, **AP-20**, 442–446.

[15] Faché, N., Van Hese, J., and De Zutter, D. (1989). Generalized space domain Green's dyadic for multilayered media with special application to microwave interconnections. *Journal of Electromagnetic Waves and Applications*, **3**, 7, 651–669.

[16] Beyne, L. and De Zutter, D. (1988). Green's function of layered lossy media with special application to microstrip antennas. *IEEE Transactions on Microwave Theory and Techniques*, **MTT-36**, 5, 875–881.

[17] Sphicopoulos, T., Theodoris, V., and Gardiol, F. (1985). Dyadic Green's function for the electromagnetic field in multilayered isotropic media: an operator approach. *Proceedings of the Institute of Electrical Engineers*, **132**, H, 5, 329–334.

[18] Wang, J. (1985). General method for the computation of radiation in stratified media. *Proceedings of the Institute of Electrical Engineers*, **132**, H, 1, 58–62.

[19] Sugita, E. and Ibaragi, O. (1979). Reliable multiwire circuits with small gauge wires. *IEEE Transactions on Components, Hybrids and Manufacturing Technology*, **CHMT-2**, 532–536.

[20] Messner, G. (1987). Cost-density analysis of interconnections. *IEEE Transactions on Components, Hybrids and Manufacturing Technology*, **CHMT-10**, 143–151.

[21] Shibata, H. and Terakado, R. (1984). Characteristics of transmission lines with a single wire for a multiwire circuit board. *IEEE Transactions on Microwave Theory and Techniques*, **MTT-32**, 4, 360–364.

[22] Wei, C., Harrington, R. F., Mautz, J. R., and Sarkar, T. (1984). Multiconductor transmission lines in multilayered dielectric media. *IEEE Transactions on Microwave Theory and Techniques*, **MTT-32**, 4, 439–444.

[23] Delbare, W. and De Zutter, D. (1989). Space-domain Green's function approach to the capacitance calculation of multiconductor lines in multilayered dielectrics with improved surface charge modelling. *IEEE Transactions on Microwave Theory and Techniques*, **MTT-37**, 10, 1562–1568.

[24] Olyslager, F., Faché, N., and De Zutter, D. (1991). New fast and accurate line parameter calculation of general multiconductor transmission lines in multilayered media. *IEEE Transactions on Microwave Theory and Techniques*, **MTT-39**, 6, 901–909.

[25] Stanghan, C. J. and Macdonal, B. M. (1985). Electric characterization of packages for high-speed integrated circuits. *IEEE Transactions on Components, Hybrids and Manufacturing Technology*, **CHMT-8**, 468–473.

[26] Faché, N. and De Zutter, D. (1989). Full-wave analysis of a perfectly conducting wire transmission line in a double layered conductor-backed medium. *IEEE Transactions on Microwave Theory and Techniques*, **MTT-37**, 3, 512–518.

[27] Faché, N., Olyslager, F., and De Zutter, D. (1991). Full-wave analysis of coupled perfectly conducting wires in a multilayered medium. *IEEE Transactions on Microwave Theory and Techniques*, **MTT-39**, 4, 673–681.

[28] Michalski, K. A. and Zheng, D. (1989). Rigorous analysis of open microstrip lines of arbitrary cross section in bound and leaky regimes. *IEEE Transactions on Microwave Theory and Techniques*, **MTT-37**, 12, 2005–2010.

[29] Olyslager, F. and De Zutter, D. (1991). New boundary integral equation for waveguides with arbitrary cross section embedded in a multilayered dielectric. In *1991 International Symposium Digest IEEE Antennas and Propagation*, Volume II, pp. 866–869, London, Ontario.

[30] Olyslager, F., Blomme, K., and De Zutter, D. (1992). Full-wave eigenmode determination of propagation constants and impedances of coupled polygonal conductors in multilayered media. In *Proceedings of the 1992 URSI International Symposium on Electromagnetic Theory*, pp. 400–402, Sydney, Australia.

[31] Olyslager, F., De Zutter, D., and Blomme, K. (1993). Rigorous analysis of the propagation characteristic of general lossless and lossy multiconductor transmission lines in multilayered media. *IEEE Transactions on Microwave Theory and Techniques.* (In press.)

6
SPECTRAL FIELD CALCULATIONS IN A MULTILAYERED MEDIUM

6.1 Introduction

The transmission line structures analysed in Chapters 7–9 consist of conductors embedded in a multilayered medium. We distinguish three types of media of which the cross-sections are shown in Fig. 6.1: a closed type (Fig. 6.1(a)), a semi-open type (Fig. 6.1(b)), and an open type (Fig. 6.1(c)). In the closed type the layers are sandwiched between two ground planes. In the semi-open type the medium is placed on top of a ground plane and the top layer is semi-infinite. In the open type both outermost layers are semi-infinite.

Each type of medium consists of an arbitrary number of layers. The layers are numbered from bottom to top and an arbitrary layer is referred to by i, where $i = 1, \ldots, L$ with L the total number of layers. Each layer is homogeneous and isotropic and is characterized by a complex permittivity ε_i and permeability μ_i. The thickness of layer i is d_i.

The right-handed coordinate system used in this and the following chapters is chosen in the following way. The x-axis coincides with the longitudinal direction of the structure, that is, the direction perpendicular to the cross-section. The y-axis is parallel to the layers and located in the cross-section. Lastly, the z-axis is perpendicular to the layers. The interface between two adjacent layers i and $i + 1$ has a transverse position denoted by $z = z_{i,i+1}$. Each layer i has a bottom and a top interface with transverse coordinates $z = z_{i-1,i}$ and $z = z_{i,i+1}$, respectively ($i = 1, \ldots, L$). By convention, for a semi-infinite outermost layer the top and bottom interfaces coincide. Ground planes are considered as interfaces. A ground plane in layer 1 or L has a transverse coordinate $z = z_{0,1}$ or $z = z_{L,L+1}$, respectively.

The full-wave analyses presented in the next three chapters aim at the characterization of the modes propagated by the interconnection structure under study. The modal fields $[\mathbf{e}(x, y, z), \mathbf{h}(x, y, z)]$ can be written in the following general form:

$$\mathbf{e}(x, y, z) = \mathbf{E}(y, z)\, e^{-j\beta x}$$
$$\mathbf{h}(x, y, z) = \mathbf{H}(y, z)\, e^{-j\beta x}. \tag{6.1}$$

We assume a sinusoidal time variation $e^{j\omega t}$. The propagation constant β is real for propagating modes in lossless structure and becomes complex for radiating modes or modes in lossy structures. The minus sign in the exponential wave term of (6.1) indicates that the mode is propagating in the

6.1 INTRODUCTION

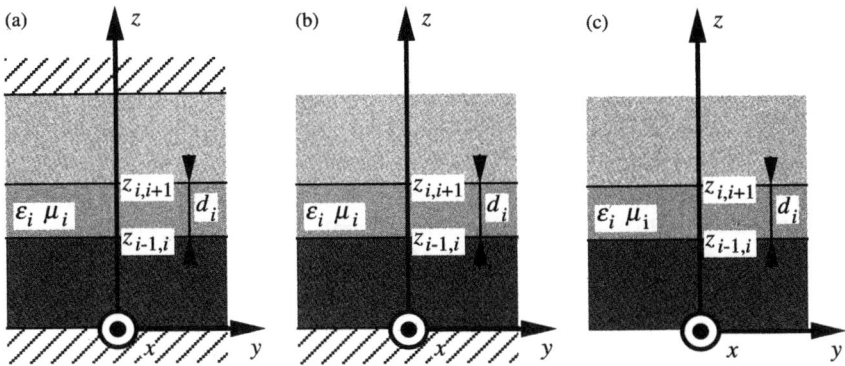

Fig. 6.1. Three types of multilayered media: (a) a closed type; (b) a semi-open type; (c) an open type.

positive longitudinal direction. The electric field $\mathbf{E}(y, z)$ and the magnetic field $\mathbf{H}(y, z)$ represent the mode in sinusoidal regime. Both fields are a solution to the two-dimensional Helmholtz equation

$$\nabla_t^2 \mathbf{E}(y, z) + \gamma^2 \mathbf{E}(y, z) = \mathbf{0}$$
$$\nabla_t^2 \mathbf{H}(y, z) + \gamma^2 \mathbf{H}(y, z) = \mathbf{0} \qquad (6.2)$$

with $\gamma^2 = \omega^2 \varepsilon \mu - \beta^2$. The gradient operator is given by

$$\nabla_t = \frac{\partial}{\partial y} \mathbf{1}_y + \frac{\partial}{\partial z} \mathbf{1}_z. \qquad (6.3)$$

The Helmholtz equation in (6.2) cannot be solved analytically. However, the structure is translationally invariant in the y-direction and therefore it is beneficial to Fourier transform the fields along this direction. This operation is often called a spectral transformation of the fields; it transforms the fields from the space to the spectral domain. The physical meaning of and the mathematics behind this operation will be addressed in more detail later in this chapter. For now the transformation of the Helmholtz equation is of importance. The two-dimensional, second-order differential equation in (6.2) is transformed into a one-dimensional, second-order differential equation:

$$\frac{d^2}{dz^2} \mathbf{E}(k_y, z) + (\gamma^2 - k_y^2) \mathbf{E}(k_y, z) = \mathbf{0}$$
$$\frac{d^2}{dz^2} \mathbf{H}(k_y, z) + (\gamma^2 - k_y^2) \mathbf{H}(k_y, z) = \mathbf{0}. \qquad (6.4)$$

The spectral fields **E** and **H** depend on the spectral variable k_y (instead of y in the space domain) and on the z-coordinate. The spectral Helmholtz equation can be solved analytically in each layer of the structure. The spectral fields everywhere in the structure follow from the analytical solution of the Helmholtz equation, the relation between the electric and magnetic fields in each layer through the Maxwell equations, the boundary conditions at all the interfaces (including ground planes and semi-infinite layers), and the transformed excitation of the fields. An efficient chain matrix formalism will be discussed in this chapter to find the spectral fields everywhere in the multilayered structure. The inverse Fourier transformation yields the spatial fields.

The spectral domain technique has become a very powerful tool for solving a wide variety of electromagnetic field problems. This chapter presents the spectral domain technique with application to planar stratified media. The presentation here is an extension of the work of Mittra and Itoh who where the first to use the spectral domain technique for microstrip structures [1]. The original technique has been improved and used for more complicated structures [2].

In the first part of this chapter we calculate the spectral fields starting from the fields in the space domain. We will show that arbitrary fields in the spectral domain can always be expressed in terms of TE and TM modes. These modes propagate in the z-direction. Next, we present a chain matrix formalism which solves for the TE and TM modes of an arbitrary multi-layered structure. In the second part we calculate the spatial fields as the inverse Fourier transforms of the spectral fields. We will address in more detail two important problems occurring in this back transformation to the spatial domain. The first is the modification of the integration path of the inverse Fourier integral in order to cope with possible poles appearing in the integrand. The second problem is the asymptotic behaviour of the integrand, that is, its behaviour for large absolute values of k_y. Finally, in the third part we will show how the power propagated in a sourceless layer can be calculated efficiently in the spectral domain.

The spectral technique presented in this chapter is also applicable to multilayered media bounded by sidewalls and in particular to media enclosed in a box. The Fourier integrals are then replaced by Fourier series.

6.2 The electromagnetic field in the spectral domain

This section constitutes the major part of this chapter. We start with a description of the electromagnetic field in a sourceless layer of the structure. Next we Fourier transform the field equations to the spectral domain. This step will reduce the complexity of the equations governing the field behaviour

6.2 ELECTROMAGNETIC FIELD IN THE SPECTRAL DOMAIN

and enables us to solve the equations analytically in terms of TE and TM modes. Next, we present a general chain matrix technique for the calculation of the spectral fields in a layered medium. The starting point is the transformation formulas for the spectral fields within a single layer. We conclude this section with some numerical considerations to be taken into account for the inverse Fourier transformation of the spectral fields.

6.2.1 *The spatial electromagnetic field in a sourceless layer*

The electromagnetic field in a sourceless layer i is described by the Maxwell equations:

$$\begin{aligned} \nabla \times \mathbf{E}_i &= -j\omega\mu_i\mathbf{H}_i \\ \nabla \times \mathbf{H}_i &= j\omega\varepsilon_i\mathbf{E}_i \\ \nabla \cdot \mathbf{D}_i &= 0 \\ \nabla \cdot \mathbf{B}_i &= 0 \end{aligned} \tag{6.5}$$

with the gradient operator given by

$$\nabla = -j\beta\mathbf{1}_x + \frac{\partial}{\partial y}\mathbf{1}_y + \frac{\partial}{\partial z}\mathbf{1}_z. \tag{6.6}$$

The longitudinal variation of a mode is already taken into account as the partial derivative with respect to x is replaced by $-j\beta$. From equations (6.5) one can easily find the wave equation for the electric and magnetic fields:

$$\begin{aligned} \nabla_t^2\mathbf{E}_i(y,z) + \gamma^2\mathbf{E}_i(y,z) &= 0 \\ \nabla_t^2\mathbf{H}_i(y,z) + \gamma^2\mathbf{H}_i(y,z) &= 0 \end{aligned} \tag{6.7}$$

with the gradient operator ∇_t defined in (6.3). These equations are called the Helmholtz equations. They are, as was already mentioned in the introduction, two-dimensional second-order differential equations in the space coordinates y and z. The x dependence does not occur in these equations as we assumed a longitudinal variation of $e^{-j\beta x}$ for the modal fields.

Besides the Maxwell equations we also need the boundary conditions for the electromagnetic field at all the interfaces. For these conditions we distinguish three types of interfaces: the interface between two adjacent layers, a ground plane, and a semi-infinite outermost layer. The different boundary conditions will now be discussed successively.

The boundary conditions at the interface between the layers i and $i+1$, that is, at the plane $z = z_{i,i+1}$, are given by

$$\begin{aligned} -\mathbf{1}_z \times \mathbf{E}_i(y, z = z_{i,i+1}) + \mathbf{1}_z \times \mathbf{E}_{i+1}(y, z = z_{i,i+1}) &= 0 \\ -\mathbf{1}_z \times \mathbf{H}_i(y, z = z_{i,i+1}) + \mathbf{1}_z \times \mathbf{H}_{i+1}(y, z = z_{i,i+1}) &= 0 \end{aligned} \tag{6.8}$$

and

$$-\mathbf{1}_z \cdot \mathbf{D}_i(y, z = z_{i,i+1}) + \mathbf{1}_z \cdot \mathbf{D}_{i+1}(y, z = z_{i,i+1}) = 0$$
$$-\mathbf{1}_z \cdot \mathbf{B}_i(y, z = z_{i,i+1}) + \mathbf{1}_z \cdot \mathbf{B}_{i+1}(y, z = z_{i,i+1}) = 0.$$
(6.9)

The first set (6.8) expresses the continuity of the tangential components (the components located in the interface plane) of the electric and magnetic fields. The second set (6.9) expresses the continuity of the normal components of the **D** and **B** fields.

In general, a ground plane can be perfectly conducting or can have a finite conductivity. In the sequel we will assume perfect conductivity. Finite conductivity can easily be incorporated by modelling the ground plane as a semi-infinite lossy layer. For a very good conductor, the skin effect can be accounted for by introducing a frequency-dependent surface impedance. However, all the simulations in the next three chapters assume perfectly conducting ground planes. The boundary conditions at a perfectly conducting ground plane require the tangential electric field to be zero:

$$\mathbf{1}_z \times \mathbf{E}_1(y, z = z_{0,1}) = \mathbf{0}$$
$$\mathbf{1}_z \times \mathbf{E}_L(y, z = z_{L,L+1}) = \mathbf{0}.$$
(6.10)

For a semi-infinite layer only outgoing waves can exist. Incoming waves from infinity must be excluded. This follows from the causality principle [3]. We will take a closer look at this boundary condition when dealing with the calculation of the spectral fields in a multilayered structure (see Section 6.2.3).

6.2.2 *The spectral transformation of the electromagnetic field*

The translational invariance in the y-direction of the multilayered media under study can be utilized in the solution technique for the fields. Indeed, if we Fourier transform the spatial fields governed by the Maxwell equations (6.5), the resulting spectral fields are a solution to a one-dimensional second-order differential equation which can be solved analytically. We discuss this spectral transformation in this section.

The Fourier transformation of a function $f(y)$ and its inverse are defined by the following relations:

$$f(k_y) = \frac{1}{2\pi} \int_{-\infty}^{+\infty} f(y) \, e^{jk_y y} \, dy$$
(6.11)

and

$$f(y) = \int_{-\infty}^{+\infty} f(k_y) \, e^{-jk_y y} \, dk_y.$$
(6.12)

6.2 ELECTROMAGNETIC FIELD IN THE SPECTRAL DOMAIN

Because the spatial and spectral calculations are kept separate, in this and the following chapters no new notation for the spectral fields is required as there will be no danger for confusion. A function and its transform are distinguished by their argument.

The inverse Fourier transformation in (6.12) shows that the spectral transformation decomposes the mode of the interconnection structure under study in waves of the form $e^{-j(\beta x + k_y y)}$ where k_y is an arbitrary real number. The modes of each layer can be written as $e^{-j(k_x x + k_y y)}$ where now both k_x and k_y are arbitrary real numbers [3]. The variation in the z-direction of the layer modes $e^{-j(\beta x + k_y y)}$ remains undetermined and follows from the spectral field solution. From this we can conclude that the spatial modal fields are written as an integral over a subclass ($k_x = \beta$) of the modes of the layers of the structure. The longitudinal direction for these modes is the z-direction.

In practical circuit design, one is often interested in the modes of a transmission line or in the S-matrix of a general planar structure enclosed in a box such as a package. In this case the layers no longer extend to infinity and sidewalls are present. The spectral technique is still applicable but now the spatial fields are written as a Fourier series instead of a Fourier integral of the spectral fields.

Fourier transformation of the Maxwell equations (6.5) yields the equations governing the behaviour of the spectral fields in layer i:

$$-j\mathbf{k} \times \mathbf{E}_i + \mathbf{1}_z \times \frac{d\mathbf{E}_i}{dz} = -j\omega\mu_i \mathbf{H}_i$$

$$-j\mathbf{k} \times \mathbf{H}_i + \mathbf{1}_z \times \frac{d\mathbf{H}_i}{dz} = j\omega\varepsilon_i \mathbf{E}_i$$

$$-j\mathbf{k} \cdot \mathbf{E}_i + \frac{dE_{z,i}}{dz} = 0$$

$$-j\mathbf{k} \cdot \mathbf{H}_i + \frac{dH_{z,i}}{dz} = 0.$$

(6.13)

The k-vector is given by $\mathbf{k} = \beta \mathbf{1}_x + k_y \mathbf{1}_y$. As can easily be seen from (6.8) and (6.9) the boundary conditions at the interface between two layers and at a perfectly conducting ground plane remain unchanged, that is, we have only to replace y by k_y in the arguments.

6.2.3 General representation of the spectral fields: TE and TM modes

From the spectral Maxwell equations follow the one-dimensional Helmholtz equations governing the behaviour of the spectral electric and magnetic fields:

$$\frac{d^2 \mathbf{E}_i}{dz^2} - \Gamma_i^2 \mathbf{E}_i = 0$$

$$\frac{d^2 \mathbf{H}_i}{dz^2} - \Gamma_i^2 \mathbf{H}_i = 0 \tag{6.14}$$

with

$$\Gamma_i^2 = k_y^2 - \gamma_i^2. \tag{6.15}$$

The function Γ_i is defined as the square root of Γ_i^2 with positive real and hence negative imaginary part. The equations (6.14) can also be found directly from the Fourier transformation of the spatial Helmholtz equations (6.7).

The general solution to (6.14) is a sum of two exponential functions propagating in the positive and negative z-direction. For each layer we find the following form for the spectral fields:

$$\mathbf{E}_i(k_y, z) = \mathbf{A}_i \, e^{-\Gamma_i z} + \mathbf{B}_i \, e^{\Gamma_i z}$$

$$\mathbf{H}_i(k_y, z) = \mathbf{K}_i \, e^{-\Gamma_i z} + \mathbf{L}_i \, e^{\Gamma_i z}. \tag{6.16}$$

The constant vectors $\mathbf{K}_i$ and $\mathbf{L}_i$ are not independent of the vectors $\mathbf{A}_i$ and $\mathbf{B}_i$.

In order to establish the relations between the four vector coefficients in (6.16) we first introduce a new coordinate system which will replace the (x, y, z) cartesian coordinate system. In the new coordinate system an arbitrary vector $\mathbf{W}$ is characterized by three numbers W_z, W', and W'' as follows:

$$\mathbf{W} = W_z \mathbf{1}_z + \frac{W' \mathbf{k} + W''(\mathbf{1}_z \times \mathbf{k})}{k^2} \tag{6.17}$$

with $k^2 = \beta^2 + k_y^2$. The choice of this coordinate system is inspired by the scalars and vectors appearing in the Maxwell equations (6.13). For example, the divergence equations, that is, the last two equations of (6.13), contain the projection of the electric and magnetic fields onto the $\mathbf{k}$ and z-axes. In the coordinate system defined in (6.17) the z-axis, the $\mathbf{k}$-vector, and the $(\mathbf{1}_z \times \mathbf{k})$-vector determine the three coordinate axes. It is clear that the electric field components (E', E'', E_z) and magnetic field components (H', H'', H_z) are of the same form (6.16) as the electric and magnetic field vectors but with the vector coefficients replaced by scalars.

The Maxwell equations can be rewritten in terms of the electric field components and the magnetic field components in the new coordinate system. After substitution of the general solution for these field components in the rewritten Maxwell equations we finally arrive at the following

6.2 ELECTROMAGNETIC FIELD IN THE SPECTRAL DOMAIN

representation of the fields in each layer i of the structure:

$$E'_i(z, k_y, \beta) = A'_i e^{-\Gamma_i z} + B'_i e^{\Gamma_i z}$$

$$H''_i(z, k_y, \beta) = \frac{j\omega\varepsilon_i}{\Gamma_i}(A'_i e^{-\Gamma_i z} - B'_i e^{\Gamma_i z}) \quad (6.18)$$

$$E_{z,i}(z, k_y, \beta) = \frac{-j}{\Gamma_i}(A'_i e^{-\Gamma_i z} - B'_i e^{\Gamma_i z})$$

and

$$E''_i(z, k_y, \beta) = A''_i e^{-\Gamma_i z} + B''_i e^{\Gamma_i z}$$

$$H'_i(z, k_y, \beta) = \frac{\Gamma_i}{-j\omega\mu_i}(A''_i e^{-\Gamma_i z} - B''_i e^{\Gamma_i z}) \quad (6.19)$$

$$H_{z,i}(z, k_y, \beta) = \frac{1}{\omega\mu_i}(A''_i e^{-\Gamma_i z} + B''_i e^{\Gamma_i z}).$$

The derivation of (6.18) and (6.19) is left to the reader as an exercise. From (6.18) and (6.19) we can conclude that the spectral representation of the fields separates into two sets of decoupled equations: one set for E', H'', and E_z, and a second set for E'', H', and H_z. The field components in (6.18) and (6.19) describe the amplitudes of the electric and magnetic fields associated with a mode $e^{-j(\beta x + k_y y)}$. The two exponential terms in each component in (6.18) and (6.19) describe the forward and backward propagation of the mode $e^{-j(\beta x + k_y y)}$ in the z-direction. The z-direction is the longitudinal direction for the modes of the layers. The first set has only a z-directed component of the electric field while the second set has only a z-directed component of the magnetic field. Therefore the first set is called a TM mode while the second set is called a TE mode.

In the representation (6.18) and (6.19) of the spectral fields four yet to be determined coefficients occur: A' and B' for the TM mode and A'' and B'' for the TE mode. These coefficients follow from the boundary conditions of the electromagnetic field.

At the interface between the layers i and $i+1$ we have the continuity of the tangential electric and magnetic fields:

$$E'_i(k_y, z = z_{i,i+1}) = E'_{i+1}(k_y, z = z_{i,i+1})$$
$$E''_i(k_y, z = z_{i,i+1}) = E''_{i+1}(k_y, z = z_{i,i+1}) \quad (6.20)$$

and

$$H'_i(k_y, z = z_{i,i+1}) = H'_{i+1}(k_y, z = z_{i,i+1})$$
$$H''_i(k_y, z = z_{i,i+1}) = H''_{i+1}(k_y, z = z_{i,i+1}). \quad (6.21)$$

It can be shown that the boundary conditions (6.9) for the normal field components do not introduce new relations between the field components. If, in the following, we refer to the boundary conditions at an interface we assume the tangential boundary conditions (6.20) and (6.21).

At a perfectly conducting ground plane in either layer 1 or layer L the zero tangential electric field in the new coordinate system is now expressed as follows:

$$E'_1(k_y, z = z_{0,1}) = E''_1(k_y, z = z_{0,1}) = 0$$
$$E'_L(k_y, z = z_{L,L+1}) = E''_L(k_y, z = z_{L,L+1}) = 0. \tag{6.22}$$

In the case of a semi-infinite outermost layer we require that no waves come in from infinity. Taking our sign convention for Γ_i into account, this boundary condition implies that for layer 1 the wave $e^{-\Gamma_1 z}$ vanishes in (6.18) and (6.19) while for layer L the wave $e^{\Gamma_L z}$ has zero amplitude.

6.2.4 The transmission line cascade model for the TE and TM modes

From the general spectral representation (6.18) and (6.19) it follows that the couples (E', H'') and (E'', H') in each layer behave as voltage and current pairs across a homogeneous transmission line. The classical representation of the voltage across and the current in a transmission line with complex propagation constant γ and characteristic impedance Z_0 is given by

$$V(z) = V_1 e^{-\gamma z} + V_2 e^{\gamma z}$$
$$I(z) = \frac{1}{Z_0}(V_1 e^{-\gamma z} - V_2 e^{\gamma z}). \tag{6.23}$$

Identification of (6.23) with (6.18) and (6.19) shows that both the TE and TM mode propagate with a propagation constant Γ_i in layer i but with a different characteristic impedance. For the TM mode we have

$$Z_{\text{TM},i}(k_y) = \frac{\Gamma_i}{j\omega\varepsilon_i} \tag{6.24}$$

while for the TE mode the characteristic impedance in layer i is given by

$$Z_{\text{TE},i}(k_y) = \frac{-j\omega\mu_i}{\Gamma_i}. \tag{6.25}$$

The z-directed components of the electric and magnetic field follow from the tangential components and are therefore dependent components. We will calculate only the tangential components.

The transmission lines associated with both (E', H'') and (E'', H') can now

6.2 ELECTROMAGNETIC FIELD IN THE SPECTRAL DOMAIN

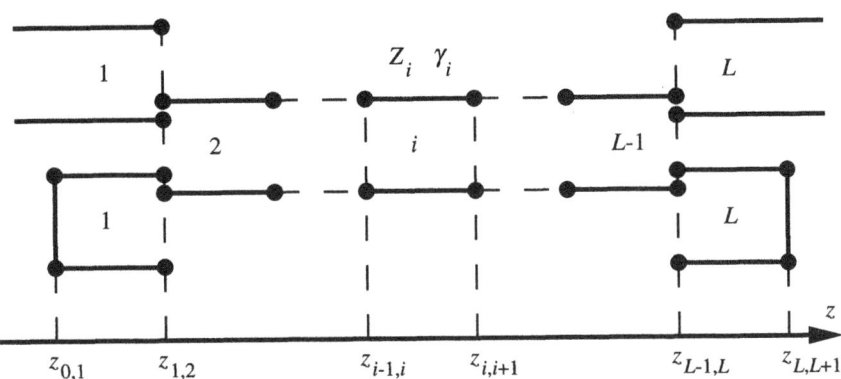

Fig. 6.2. Cascade of L transmission lines. The outermost layers are either semi-infinite or short circuited.

be interconnected since the continuity of the tangential electric and magnetic fields at the interface between two layers corresponds to the continuity of the voltage and the current at the interface between two adjacent transmission lines. Furthermore, a perfectly conducting ground plane is translated to a short circuit as the electric field, either E' or E'', equals zero. A semi-infinite layer is modelled as a semi-infinite transmission line. The line obeys the same 'no incoming wave' boundary condition as the semi-infinite layer.

We can now conclude that the TE and the TM modes are solutions to a transmission line cascade. Such a cascade is sketched in Fig. 6.2.

Before solving the TE and TM cascade we still need the excitations or the sources of the cascade. These excitations depend on the electromagnetic field we want to calculate and therefore also on the full-wave analysis technique used. In the following three chapters we will discuss the excitations in more detail. In this chapter we assume that for both the TE and TM cascade the excitation consists of an arbitrary voltage source and current source located at the interface between two layers e and $e + 1$ ($e = 1, \ldots, L$); see Fig. 6.3. Such an interface can also be a fictitious separation of two parts of a single homogeneous layer. It will turn out that all excitations considered in the next chapters can be represented in terms of such voltage and current sources. In the next section we will derive the transformation formulas for the TE and TM fields in each layer and between adjacent layers.

6.2.5 Transformation formulas in the TE and TM cascade

The solution technique for the TE and TM cascade is similar and therefore we first introduce a generic notation for both cascades. Each line in the generic cascade has an impedance Z_i and a propagation constant Γ_i. The

122   6 SPECTRAL FIELD CALCULATIONS

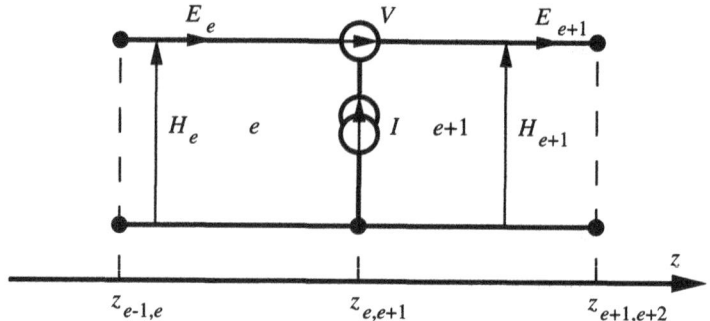

Fig. 6.3. Voltage–current excitation of the TE or TM cascade.

field pair in a transmission line i is (E_i, H_i). The excitation is a voltage V and a current I located at a position $z = z_{e,e+1}$. This generic notation was introduced in Fig. 6.3.

The building blocks in the solution technique for the cascade are the transformation formulas which transform the field pair at either the top interface or the bottom interface to an arbitrary position z in that layer. In order to obtain these transformation formulas we introduce two new representations for the couple (E_i, H_i) in terms of hyperbolic functions of the transverse position z instead of the exponential functions used before.

In a first representation the hyperbolic functions are referred to the bottom interface. In matrix notation we have for layer i:

$$\begin{bmatrix} E_i(z) \\ H_i(z) \end{bmatrix} = \begin{bmatrix} \cosh \alpha_i^u(z) & -Z_i \sinh \alpha_i^u(z) \\ (-1/Z_i) \sinh \alpha_i^u(z) & \cosh \alpha_i^u(z) \end{bmatrix} \begin{bmatrix} E_i(z = z_{i-1,i}) \\ H_i(z = z_{i-1,i}) \end{bmatrix} \quad (6.26)$$

with

$$\alpha_i^u(z) = \Gamma_i(z - z_{i-1,i}). \quad (6.27)$$

In (6.26) we introduced a set of wave coefficients for the hyperbolic waves. These coefficients are the electric and magnetic fields at the bottom interface of layer i. The transformation (6.26) transforms the field pair (E_i, H_i) from the bottom interface to an arbitrary position z in the layer i and is therefore called an up transformation. This explains the superscript 'u' in the argument of the hyperbolic functions.

In the second representation the fields are transformed from the top interface of layer i to an arbitrary position in that layer. This down transformation is given by

$$\begin{bmatrix} E_i(z) \\ H_i(z) \end{bmatrix} = \begin{bmatrix} \cosh \alpha_i^d(z) & Z_i \sinh \alpha_i^d(z) \\ (1/Z_i) \sinh \alpha_i^d(z) & \cosh \alpha_i^d(z) \end{bmatrix} \begin{bmatrix} E_i(z = z_{i,i+1}) \\ H_i(z = z_{i,i+1}) \end{bmatrix} \quad (6.28)$$

6.2 ELECTROMAGNETIC FIELD IN THE SPECTRAL DOMAIN

with

$$\alpha_i^d(z) = \Gamma_i(z_{i,i+1} - z). \tag{6.29}$$

We will use a shorthand notation for the field column vector and the transformation matrices. The fields at an arbitrary position z in layer i are denoted by the column vector $\mathbf{C}_i(z)$:

$$\mathbf{C}_i(z) = \begin{bmatrix} E_i(z) \\ H_i(z) \end{bmatrix}. \tag{6.30}$$

The up and down transformation matrices are written as

$$\mathbf{M}_i^u(z) = \begin{bmatrix} \cosh \alpha_i^u(z) & -Z_i \sinh \alpha_i^u(z) \\ (-1/Z_i) \sinh \alpha_i^u(z) & \cosh \alpha_i^u(z) \end{bmatrix} \tag{6.31}$$

and

$$\mathbf{M}_i^d(z) = \begin{bmatrix} \cosh \alpha_i^d(z) & Z_i \sinh \alpha_i^d(z) \\ (1/Z_i) \sinh \alpha_i^d(z) & \cosh \alpha_i^d(z) \end{bmatrix}. \tag{6.32}$$

If we omit the z argument we assume a transformation over the total layer i, that is, from $z = z_{i-1,i}$ to $z = z_{i,i+1}$ for an up transformation and from $z = z_{i,i+1}$ to $z = z_{i-1,i}$ for a down transformation. For example, the matrix $\mathbf{M}_i^u$ transforms the fields from the bottom to the top interface of layer i.

The transformation formulas derived above for the fields inside a layer i can now be applied to the transformation of the fields between adjacent layers. The boundary condition at the sourceless interface between the layers i and $i + 1$ requires the electric field E and the magnetic field H to be continuous. This means that the field vector at the top interface of layer i equals the field vector at the bottom interface of layer $i + 1$. Therefore an up transformation of the field vector from the bottom interface of layer i to the top interface of layer i yields the field vector at the bottom interface of layer $i + 1$:

$$\mathbf{C}_{i+1}(z = z_{i,i+1}) = \mathbf{M}_i^u \mathbf{C}_i(z = z_{i-1,i}). \tag{6.33}$$

Analogously, we can transform the fields at the top interface of layer i to the top interface of layer $i - 1$:

$$\mathbf{C}_{i-1}(z = z_{i-1,i}) = \mathbf{M}_i^d \mathbf{C}_i(z = z_{i,i+1}). \tag{6.34}$$

Before solving the TE and TM cascade for a particular excitation we still have to take into account the boundary conditions in the outermost layers. In these layers the boundary conditions reduce the number of unknown wave coefficients from two to one. For layer 1, the relevant field vector for further calculations is the field vector at the bottom interface. It is given by

$$\mathbf{C}_1(z = z_{0,1}) = \begin{bmatrix} -\sigma_1 Z_1 \\ 1 \end{bmatrix} H(z = z_{0,1}) = \mathbf{K}_1 H(z = z_{0,1}) \qquad (6.35)$$

where σ_1 depends on the type of layer 1. For a ground plane we have $\sigma_1 = 0$ corresponding to a zero tangential electric field, while for a semi-infinite layer we have $\sigma_1 = 1$ corresponding to only an outgoing wave in layer 1. For the top layer L, only the field vector at the top interface is of relevance for the solution of the TE and TM cascade. We have

$$\mathbf{C}_L(z = z_{L,L+1}) = \begin{bmatrix} -\sigma_L Z_L \\ 1 \end{bmatrix} H(z = z_{L,L+1}) = \mathbf{K}_L H(z = z_{L,L+1}) \qquad (6.36)$$

with $\sigma_L = 0$ for a ground plane and $\sigma_L = -1$ for a semi-infinite layer. The field vectors in the outermost layers are derived from the magnetic field. We remind the reader that in the case of a semi-infinite outermost layer the top and bottom interfaces coincide at the interface with the adjacent layer.

With the transformation formulas between layers and the relevant field vectors in the outermost layers which take the boundary condition at the outer interfaces into account, we are ready to solve the TE and TM cascade. The solution technique is presented in the next section.

6.2.6 *Solution of the TE and TM cascade*

The transformation formulas (6.33) and (6.34) will now be applied in the solution technique for the TE and TM cascade. We use a standard transmission or chain matrix technique.

As a first step we determine the fields at an arbitrary position z in layer i as a function of the fields in one of the outermost layers, either layer 1 or layer L. The transformation from layer 1 to an arbitrary position in layer i is given by

$$\mathbf{C}_i(z) = \mathbf{M}_i^u(z) \left(\prod_{k=1}^{i-1} \mathbf{M}_k^u \right) \mathbf{K}_1 H(z = z_{0,1}). \qquad (6.37)$$

The fields at z are given as a function of the magnetic field at the bottom interface of layer 1. Once the latter is determined we also know the fields at z as the transformation matrices and column vectors occurring in (6.37) are known functions. The transformation from layer L to an arbitrary position z in layer i is given by a similar expression:

$$\mathbf{C}_i(z) = \mathbf{M}_i^d(z) \left(\prod_{k=L}^{i+1} \mathbf{M}_k^d \right) \mathbf{K}_L H(z = z_{L,L+1}). \qquad (6.38)$$

6.2 ELECTROMAGNETIC FIELD IN THE SPECTRAL DOMAIN

Now the fields at z are related to the magnetic field at the top interface of layer L.

As a second step we use both transformation formulas to apply the only remaining boundary conditions in the TE and TM cascade, that is, those at the excitation interface. We assume that the excitation is located at $z = z_{e,e+1}$. The assumption that the source is located at the interface between two layers is no loss of generality as we can always introduce a fictitious interface as we mentioned before. The excitation can be described by means of a voltage source V and a current source I; see Fig. 6.3. With the transformation formula (6.37) we find the fields just below the excitation, while the transformation formula (6.38) yields the fields just above the excitation. The fields below and above the excitation exhibit a jump. The jump in E is given by the voltage V while the jump in H corresponds to the current I. The boundary conditions at the excitation are thus given by

$$\left(\prod_{k=L}^{e+1} \mathbf{M}_k^d\right) \mathbf{K}_L H(z = z_{L,L+1}) - \left(\prod_{k=1}^{e} \mathbf{M}_k^u\right) \mathbf{K}_i H(z = z_{0,1}) = \begin{bmatrix} V \\ I \end{bmatrix}. \quad (6.39)$$

The unknown scalars in these boundary conditions are the magnetic fields at the bottom interface of layer 1 and the top interface of layer L. Solution of (6.39) expresses these two unknowns in terms of the voltage V and the current I. The transformation formulas (6.37) and (6.38) then give the spectral fields everywhere in the cascade.

6.2.7 Numerical considerations

Before moving on to the calculation of the electromagnetic field in the space domain, we first discuss three important aspects of the solution of the TE and TM cascade with a view to its numerical integration in the inverse Fourier transformation.

The first aspect of the solution is the unbounded nature of the hyperbolic functions (as a function of k_y) which occur in the elements of the transformation matrices used above. Therefore the transformation formulas in their present form are not suited to numerical evaluation. A solution to this problem consists of a reformulation of the chain matrix technique in terms of transformation matrices which have no unbounded exponential terms in their elements.

The second aspect of practical importance is the asymptotic behaviour of the solution of the TE and TM cascade, that is, the solution for large absolute values of the spectral variable k_y. The inverse Fourier transformation of the asymptotic spectral fields determines the spatial fields in the vicinity of the excitation. These spatial fields often become infinite at the excitation (for example, near a line current source). A separate (analytical) calculation of the asymptotic contribution is required in order to determine the spatial fields accurately. The contribution of the asymptotic part can be calculated analytically for multilayered media.

The third aspect of the solution of the cascades that we will address is the existence of poles in the integrand of the inverse Fourier integrals. Each pole corresponds to a solution of the sourceless TE or TM cascade and gives rise to surface-wave modes in the structure. In order to take these poles correctly into account we have to modify the integration path of the inverse Fourier integral. How this is done will be discussed in the Section 6.3 dealing with the electromagnetic field in the space domain.

In the following three sections we will discuss the three above-mentioned numerical problems in more detail.

6.2.7.1 *Bounded transformation matrices* All the elements of the transformation matrices in (6.39) contain exponentially increasing functions of the spectral variable k_y. However, we can rewrite the boundary condition (6.39) at the excitation with matrices which contain only decreasing exponential functions for increasing values of the spectral variable k_y. We will now discuss how this is implemented.

In order to remove the unbounded exponential terms we introduce new transformation matrices. For the up transformation the original transformation matrix $\mathbf{M}_i^u$ can be written as the product of an increasing exponential term and a new matrix, denoted $\mathbf{N}_i^u$, which contains only bounded exponential functions in each matrix element:

$$\mathbf{M}_i^u = e^{\Gamma_i d_i} \mathbf{N}_i^u \qquad (6.40)$$

with $\mathbf{N}_i^u$ given by

$$\mathbf{N}_i^u = \begin{bmatrix} \dfrac{1 + e^{-2\Gamma_i d_i}}{2} & -Z_i \dfrac{(1 - e^{-2\Gamma_i d_i})}{2} \\ \dfrac{-(1 - e^{-2\Gamma_i d_i})}{2Z_i} & \dfrac{1 + e^{-2\Gamma_i d_i}}{2} \end{bmatrix}. \qquad (6.41)$$

For the down transformation the new matrix $\mathbf{N}_i^d$ is related to the original matrix $\mathbf{M}_i^d$ in the following way:

$$\mathbf{M}_i^d = e^{\Gamma_i d_i} \mathbf{N}_i^d \qquad (6.42)$$

with $\mathbf{N}_i^d$ given by

$$\mathbf{N}_i^d = \begin{bmatrix} \dfrac{1 + e^{-2\Gamma_i d_i}}{2} & Z_i \dfrac{(1 - e^{-2\Gamma_i d_i})}{2} \\ \dfrac{1 - e^{-2\Gamma_i d_i}}{2Z_i} & \dfrac{1 + e^{-2\Gamma_i d_i}}{2} \end{bmatrix}. \qquad (6.43)$$

6.2 ELECTROMAGNETIC FIELD IN THE SPECTRAL DOMAIN

The boundary condition at the excitation can be expressed in terms of the new matrices:

$$\left(\prod_{k=L}^{e+1} e^{\Gamma_k d_k}\right)\left(\prod_{k=L}^{e+1} \mathbf{N}_k^d\right) \mathbf{K}_L H(z = z_{L,L+1})$$

$$- \left(\prod_{k=1}^{e} e^{\Gamma_k d_k}\right)\left(\prod_{k=1}^{e} \mathbf{N}_k^u\right) \mathbf{K}_1 H(z = z_{0,1}) = \begin{bmatrix} V \\ I \end{bmatrix}. \quad (6.44)$$

This expression still contains two products of exponential terms as factors placed in front of the two products of $\mathbf{N}$ matrices. However, they can be taken together with the magnetic field at the bottom interface of layer 1 and at the top interface of layer L resulting in new unknown wave coefficients R_1 and R_L, respectively:

$$R_1 = \left(\prod_{k=1}^{e} e^{\Gamma_i d_i}\right) H(z = z_{0,1})$$

$$R_L = \left(\prod_{k=L}^{e+1} e^{\Gamma_i d_i}\right) H(z = z_{L,L+1}). \quad (6.45)$$

The boundary condition (6.44) is now expressed in terms of bounded matrices and no unbounded exponential terms occur:

$$\left(\prod_{k=L}^{e+1} \mathbf{N}_k^d\right) \mathbf{K}_L R_L - \left(\prod_{k=1}^{e} \mathbf{N}_k^u\right) \mathbf{K}_1 R_1 = \begin{bmatrix} V \\ I \end{bmatrix}. \quad (6.46)$$

This set of two scalar equations can be solved numerically for any value of the spectral variable k_y and yields the new wave coefficients R_1 and R_L. Once we have these coefficients we can derive the field column vector at an arbitrary position in the cascade. For a position z in layer i ($i \leq e$) located below the excitation we have

$$\mathbf{C}_i(z) = e^{-\Gamma_i(z_{i,i+1}-z)}\left(\prod_{k=e}^{i+1} e^{-\Gamma_k d_k}\right) \mathbf{N}_i^u(z)\left(\prod_{k=1}^{i-1} \mathbf{N}_k^u\right) \mathbf{K}_1 R_1 \quad (6.47)$$

while above the excitation we find at a position z in layer i ($i \geq e+1$):

$$\mathbf{C}_i(z) = e^{-\Gamma_i(z-z_{i-1,i})}\left(\prod_{k=e+1}^{i-1} e^{-\Gamma_k d_k}\right) \mathbf{N}_i^d(z)\left(\prod_{k=L}^{i+1} \mathbf{N}_k^d\right) \mathbf{K}_L R_L. \quad (6.48)$$

The z-dependent $\mathbf{N}$ matrices are defined in the following way:

$$\mathbf{M}_i^u(z) = e^{-\Gamma_i(z-z_{i-1,i})}\mathbf{N}_i^u(z)$$

$$\mathbf{M}_i^d(z) = e^{-\Gamma_i(z_{i,i+1}-z)}\mathbf{N}_i^d(z). \quad (6.49)$$

128   6 SPECTRAL FIELD CALCULATIONS

All matrices in (6.47) and (6.48) are bounded functions of the spectral variable k_y and the factors in front of the matrices consist of decreasing exponential functions for $|k_y|$ going to infinity.

6.2.7.2 *Asymptotic behaviour of the TE and TM cascades* Under the asymptotic solution of the TE and TM cascade we understand the solution for large absolute k_y values for which the decreasing exponential terms in the $\mathbf{N}_i^d$ and $\mathbf{N}_i^u$ matrices can be neglected. The asymptotic form of these matrices becomes

$$\mathbf{N}_i^d = \begin{bmatrix} \dfrac{1}{2} & \dfrac{Z_i}{2} \\ \dfrac{1}{2Z_i} & \dfrac{1}{2} \end{bmatrix} \tag{6.50}$$

and

$$\mathbf{N}_i^u = \begin{bmatrix} \dfrac{1}{2} & -\dfrac{Z_i}{2} \\ -\dfrac{1}{2Z_i} & \dfrac{1}{2} \end{bmatrix}. \tag{6.51}$$

The approximation is valid for $\Gamma_i d_i \gg 1$ ($i = 1, \ldots, L$), with d_i the thickness of layer i. Using the matrices (6.50) and (6.51) in the boundary condition (6.46) we find the asymptotic solution of the TE and TM cascade.

A physical interpretation can now be given to this asymptotic solution. This is interesting not only to get more insight into the asymptotic wave propagation in the TE and TM cascades but also to solve the cascades analytically in the asymptotic limit. For this purpose we consider in Fig. 6.4 a layer l to the left and a layer r to the right of the excitation. The excitation

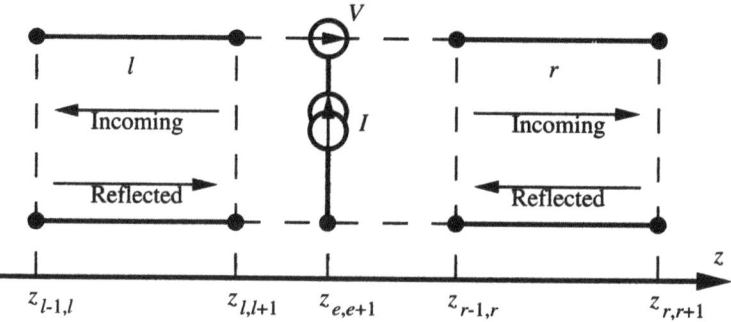

Fig. 6.4. Incoming and reflected waves in layers located to the left and to the right of the excitation.

6.2 ELECTROMAGNETIC FIELD IN THE SPECTRAL DOMAIN

itself is located at the interface between layers e and $e + 1$. In both layers l and r waves propagate in the positive and negative z-direction. With respect to the position of the excitation a wave in the positive z-direction is an incoming wave in layer r and a reflected wave in layer l. Analogously, a wave in the negative z-direction is an incoming wave in layer l and a reflected wave in layer r.

The fields in layer r are found in the following way. We solve the boundary conditions (6.46) at the excitation. This results in the magnetic field at the bottom interface of layer 1 and at the top interface of layer L. Next, we transform the field vector at the top interface of layer L to the top interface of layer r. From $\mathbf{C}_r(z = z_{r,r+1})$ we can calculate the fields everywhere in layer r using the transformation matrix $\mathbf{N}_r^d(z)$. In particular we can calculate the field vector $\mathbf{C}_r(z = z_{r-1,r})$ at the bottom interface of layer r:

$$\mathbf{C}_r(z = z_{r-1,r}) = e^{-\Gamma_r d_r} \mathbf{N}_r^d \mathbf{C}_r(z = z_{r,r+1}) \tag{6.52}$$

or, in full,

$$\begin{bmatrix} E(z = z_{r-1,r}) \\ H(z = z_{r-1,r}) \end{bmatrix} = e^{-\Gamma_r d_r} \begin{bmatrix} \dfrac{1}{2} & \dfrac{Z_i}{2} \\ \dfrac{1}{2Z_i} & \dfrac{1}{2} \end{bmatrix} \begin{bmatrix} E(z = z_{r,r+1}) \\ H(z = z_{r,r+1}) \end{bmatrix}. \tag{6.53}$$

The ratio of the electric to the magnetic field at the bottom interface is equal to the impedance associated with layer r:

$$\frac{E(z = z_{r-1,r})}{H(z = z_{r-1,r})} = Z_r. \tag{6.54}$$

This is also the ratio of the voltage and the current wave propagating in the positive direction on a transmission line. A wave coming in at the bottom interface of layer r travels through that layer and partially reflects and transmits at the top interface. The transmitted part propagates in layer $r + 1$ while the reflected part propagates back into layer r. The ratio (6.54) now proves that in the asymptotic limit the reflected wave becomes negligible at the bottom interface of layer r. Similar conclusions hold for a layer l to the left of the excitation.

The reasoning based on (6.54) now makes it possible to describe the asymptotic wave propagation in the cascade.

A wave departs from the source at $z = z_{e,e+1}$ in both layers e and $e + 1$. These two waves never return to the excitation as we neglect a reflected wave which has travelled through an entire layer. This also means that the excitation only 'feels' the adjacent layers e and $e + 1$. If we translate this interpretation to the spatial fields we can conclude that the behaviour of the spatial fields in the vicinity of the excitation depends only on the two layers e and $e + 1$. In other words, the dominant behaviour of the spatial fields close to the excitation is the same as if the excitation were placed at the interface between two semi-infinite layers with the same characteristics as layers e and $e + 1$.

The incoming fields in an arbitrary layer i are equal to a direct wave coming from the excitation. This wave is attenuated by the transmission coefficients of all the layer interfaces between the layer i and the excitation and by the exponential wave decay in each layer. The reflected wave in an arbitrary layer i follows from the reflection of the incoming wave at the top interface of layer $i - 1$ provided layer i is located to the left of the excitation, and at the bottom interface of layer $i + 1$ provided layer i is located to the right of the excitation.

6.2.7.3 Resonances in the TE and TM cascades Only for certain discrete values of the spectral variable k_y does a solution exist for the sourceless TE and TM cascade. These values can be found as the roots of the boundary condition equations (6.46) but with the excitations equal to zero. We have:

$$\left(\prod_{k=L}^{e+1} \mathbf{N}_k^d\right) \mathbf{K}_L R_L - \left(\prod_{k=1}^{e} \mathbf{N}_k^u\right) \mathbf{K}_1 R_1 = \mathbf{0}. \tag{6.55}$$

The roots of these coupled equations correspond to surface-wave modes of the structure. In the case of a simple microstrip the surface-wave modes correspond to modes of a symmetric dielectric slab waveguide for which the tangential electric field is zero at the symmetry plane of the slab waveguide (this corresponds to the perfectly conducting ground plane of the microstrip). The core of the slab waveguide has a thickness equal to twice that of the microstrip substrate and the material parameters of the core and the cladding are the same as those of the substrate and the top layer of the microstrip, respectively.

So far we have assumed that the spectral variable k_y is a real variable. However, in order to deal with the surface-wave modes we will extend k_y to a complex variable. For lossless structures the roots are located on the real and imaginary axes in opposite pairs. The real roots will occur as poles in the integrand of the inverse Fourier integrals. For lossy structures the roots are located in the complex plane but for small losses they are located close to either the real or the imaginary axis, so they are still important for the numerical evaluation of the inverse Fourier integrals.

6.3 The electromagnetic field in the space domain

From the solution of the TE and TM cascade we can derive the cartesian components of the spectral electromagnetic field. For example, for the electric

6.3 THE ELECTROMAGNETIC FIELD IN THE SPACE DOMAIN

field we have:

$$E_x = (\beta E' - k_y E'')/k^2$$
$$E_y = (k_y E' + \beta E'')/k^2 \tag{6.56}$$

with $k^2 = k_y^2 + \beta^2$. Inverse Fourier transformation of the components (6.56) yields the electric field in the space domain. In general a component of the spatial field is found by calculating an integral of the form

$$\int_{-\infty}^{+\infty} f(\Gamma_i, k_y) e^{-jk_y y} dk_y. \tag{6.57}$$

The function $f(\Gamma_i, k_y)$ depends on the spectral propagation constants Γ_i of all the layers (the index i is a generic notation) and the spectral variable k_y. The specific form of this function depends on the field component under study but is not relevant for the further discussion here.

In the complex k_y plane the inverse Fourier integration in (6.57) is located along the real axis. As we have seen above, poles of the function $f(\Gamma_i, k_y)$ may exist on the real axis (lossless structures) or close to the real axis (structures with small losses). Therefore a deformation of the original integration path into a new path is needed. We will now discuss this deformation.

The deformation of the integation path is based upon Cauchy's residue theorem. Before we can apply this theorem the integrand must be single valued. Felsen and Marcuvitz [3] have shown that the function $f(\Gamma_i, k_y)$ is an even function of each propagation factor Γ_i associated with a layer of finite thickness. This is not the case for a semi-infinite outermost layer and therefore the function $f(\Gamma_i, k_y)$ is a two-valued function of the propagation constant associated with such a layer. In the case of a microstrip the integrand of (6.57) is a two-valued function of the propagation constant of the top layer, in many practical cases air. We will use the microstrip to exemplify the deformation of the original integration path, that is, the real axis, into a new path C. In order to make the integrand single valued the complex k_y plane must be cut by two branch lines running from the branch points to infinity. In the case of a microstrip (with an air top layer) the branch points are given by

$$k_y = \pm j\sqrt{\beta^2 - \omega^2 \varepsilon \mu} = \pm \gamma = \pm(-\gamma' - j\gamma'') \qquad \gamma' \geq 0, \gamma'' \geq 0. \tag{6.58}$$

If we assume a complex propagation factor β the branch points are located in the first and third quadrant of the complex k_y plane. In the special case of a lossless structure the branch points are purely imaginary. A possible choice of the branch cuts is given in Fig. 6.5 where we assume complex poles.

In the derivation of the spectral fields we defined each Γ as the square

6 SPECTRAL FIELD CALCULATIONS

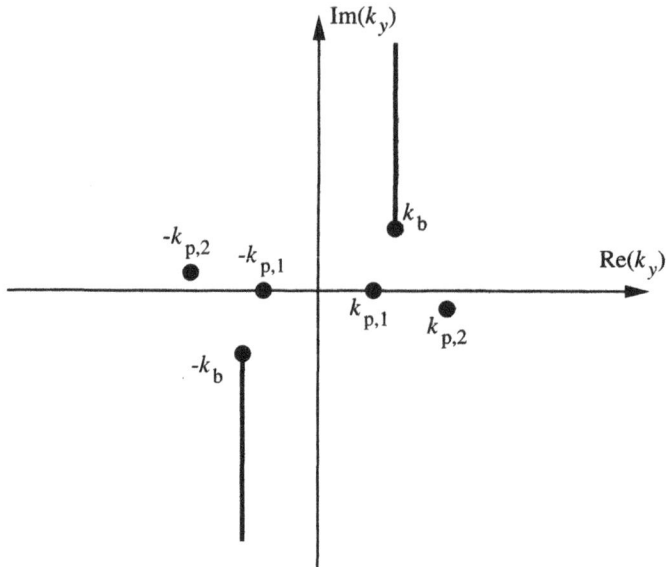

Fig. 6.5. Poles of $f(\Gamma_i, k_y)$ and branch cuts in the complex k_y plane.

root of Γ^2 with positive real and negative imaginary part. This definition is relevant only for the semi-infinite layers as we discussed above. One can easily verify that these conditions are fulfilled on the real axis for the propagation constant of the top layer of the microstrip. The integration path can be deformed anywhere into the complex k_y-plane provided it does not cross the branch cuts. For branch cuts as shown in Fig. 6.6, deformation of the path into the non-shaded (shaded) region corresponds to a representation of the field in terms of incoming (outgoing) waves. In the following, we assume that the fields are expressed in terms of incoming waves only. Let us now determine the region in the complex k_y plane for which the Γ conditions are satisfied. The conditions are satisfied when the imaginary part of Γ^2 is smaller than zero. Therefore we search for the curves along which the imaginary part of Γ^2 equals zero. If $k_y = k'_y + jk''_y$ then these curves are given by

$$k'_y k''_y = \gamma' \gamma''. \tag{6.59}$$

The expression (6.59) represents two hyperbolas running in the first and third quadrant. The branch points $\Gamma = k_b$ and $\Gamma = -k_b$ are located on the hyperbolas, as in Fig. 6.6. In the region between the two hyperbolas the imaginary part of Γ^2 is negative and therefore the condition for Γ will be satisfied.

The final step is the deformation of the integration path in order to circumvent the poles on or close to the real axis. Let us assume that $y > 0$.

6.3 THE ELECTROMAGNETIC FIELD IN THE SPACE DOMAIN 133

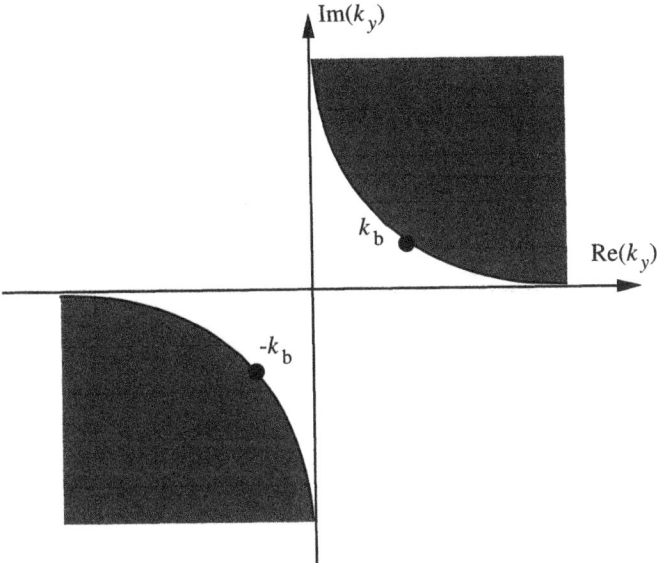

Fig. 6.6. Two hyperbolas which define the region for which Γ has a positive real and negative imaginary part. The path C must be located in the non-shaded region.

In this case the exponential factor $e^{-jk_y y}$ in the integrand of (6.57) will be exponentially decreasing in the lower half plane. Therefore we close the path C in the lower half plane. The closing contour consists of a semicircle at infinity in the lower half plane and a contour around the branch cut in the third quadrant as our integration path must not cross a branch cut. This is shown on Fig. 6.7.

The semicircle at infinity is denoted C_s while the contour around the branch cut is denoted C_b. The question which remains to be answered is which poles on the real axis must be included and which not. In order to see this we take a look at the residue of a pole at $k_y = k_p$ located on the real axis:

$$\lim_{k_y \to k_p} [f(\Gamma_i, k_y)(k_y - k_p)] e^{-jk_p y}. \tag{6.60}$$

This residue represents a travelling wave in the positive y-direction for a positive k_p value and a wave in the negative y-direction for a negative k_p value. As we assumed that $y > 0$ we must exclude the poles located on the negative real axis based on causality. The deformed path C is now shown in Fig. 6.7.

There are two ways to evaluate the integrals (6.57). Either we evaluate the integrand along C or we use Cauchy's residue theorem to express the

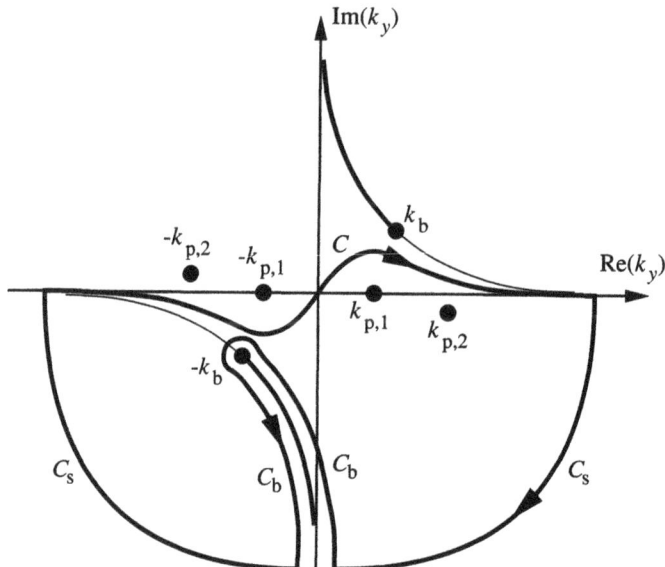

Fig. 6.7. Closing contour, deformed integration path, and branch cuts along the hyperbolas.

integrals (6.57) as a sum of the contribution of C_s and C_b and the residues of the poles enclosed by the total closed contour. The contribution of C_s remains finite and is zero only if the branch cuts are located on the hyperbolas defined in (6.59). They start in the branch points and continue along the imaginary axis. The proof is not given here; we refer the reader to [4] for more details. We note that this choice is relevant only for the second way of evaluating (6.57). For a direct evaluation of (6.57) an arbitrary choice of the branch cuts can be made.

6.4 Power propagated in a sourceless layer

The complex power P_i propagated by a sourceless layer i is given by the integral over the cross-section of the longitudinal component of the Poynting vector:

$$P_i = \tfrac{1}{2} \int_{-\infty}^{+\infty} dy \int_{z_{i-1,i}}^{z_{i,i+1}} [\mathbf{E}_i(y, z) \times \mathbf{H}_i^*(y, z)] \cdot \mathbf{1}_x \, dz. \qquad (6.61)$$

In order to calculate the power in (6.61) we need the spatial electric and magnetic fields in the cross-section. However, spatial fields are very CPU-time expensive because of the inverse Fourier transformation required for each spatial sample point. Therefore, we transform the expression (6.61) to

6.4 POWER PROPAGATED IN A SOURCELESS LAYER

the spectral domain:

$$P_i = \pi \int_{z_{i-1,i}}^{z_{i,i+1}} dz \int_{-\infty}^{+\infty} [\mathbf{E}_i(k_y, z) \times \mathbf{H}_i^*(k_y, z)] \cdot \mathbf{1}_x \, dk_y. \tag{6.62}$$

The new power expression (6.62) has two important advantages over the previous representation (6.61). Firstly, the spectral fields can be calculated very easily and are found as the solution of a TE and TM cascade as we have seen before. Secondly, the remaining space integration with respect to z can be calculated analytically as the spectral fields are written as the sum of two exponential z-directed waves. We will now work out this analytical calculation. We restrict the discussion to the power terms which combine the electric and magnetic fields of the same mode. As a first step we write the power propagated in the layer i in terms of the $(',\,'',z)$ components of E and H:

$$P_i = P_{\text{TE},i} + P_{\text{TM},i} + P_{\text{cross},i} \tag{6.63}$$

with

$$P_{\text{TE},i} = \pi\beta \int_{z_{i-1,i}}^{z_{i,i+1}} dz \int_{-\infty}^{+\infty} \frac{1}{k^2} E_i'' H_{z,i}^* \, dk_y \tag{6.64}$$

$$P_{\text{TM},i} = -\pi\beta^* \int_{z_{i-1,i}}^{z_{i,i+1}} dz \int_{-\infty}^{+\infty} \frac{1}{(k^2)^*} E_{z,i} H_i''^* \, dk_y \tag{6.65}$$

$$P_{\text{cross},i} = \pi \int_{z_{i-1,i}}^{z_{i,i+1}} dz \int_{-\infty}^{+\infty} k_y \left[\frac{1}{k^2} E_i' H_{z,i}^* - \frac{1}{(k^2)^*} E_{z,i} H_i'^* \right] dk_y. \tag{6.66}$$

The power (6.63) consists of three contributions: a TE and a TM contribution and a cross term which consists of contributions from both cascades. For the analytical integration we use the following TE and TM representation for the fields which differ slightly from the ones given in (6.18) and (6.19):

$$E_i' = A_i' e^{-\Gamma_i(z - z_{i-1,i})} + B_i' e^{-\Gamma_i(z_{i,i+1} - z)}$$

$$H_i'' = \frac{1}{Z_{\text{TM},i}} [A_i' e^{-\Gamma_i(z - z_{i-1,i})} - B_i' e^{-\Gamma_i(z_{i,i+1} - z)}] \tag{6.67}$$

$$E_{z,i} = \frac{-j}{\Gamma_i} [A_i' e^{-\Gamma_i(z - z_{i-1,i})} - B_i' e^{-\Gamma_i(z_{i,i+1} - z)}]$$

and

$$E_i'' = A_i'' e^{-\Gamma_i(z-z_{i-1,i})} + B_i'' e^{-\Gamma_i(z_{i,i+1}-z)}$$

$$H_i' = \frac{1}{Z_{TE,i}} [A_i'' e^{-\Gamma_i(z-z_{i-1,i})} - B_i'' e^{-\Gamma_i(z_{i,i+1}-z)}] \quad (6.68)$$

$$H_{z,i} = \frac{1}{\omega\mu_i} [A_i'' e^{-\Gamma_i(z-z_{i-1,i})} + B_i'' e^{-\Gamma_i(z_{i,i+1}-z)}].$$

The representations (6.67) and (6.68) contain only decreasing exponential functions. If we insert (6.67) and (6.68) in (6.64)–(6.66) the integration with respect to z can be performed analytically. The simple mathematics is left to the reader. The remaining integration with respect to k_y can be performed numerically with simple Gaussian quadrature formulas. We draw the attention of the reader to the fact that this integration must be performed along the real k_y-axis. Integration along another complex path leads to wrong results. This is a consequence of the presence of the complex conjugate operator which is not holomorphic.

References

[1] Itoh, T. and Mittra, R. (1973). Spectral-domain approach for calculating the dispersion characteristics of microstrip lines. *IEEE Transactions on Microwave Theory and Techniques*, **MTT-21**, 7, 496–499.
[2] Jansen, R. H. (1985). The spectral domain approach for microwave integrated circuits. *IEEE Transactions on Microwave Theory and Techniques*, **MTT-33**, 10, 1043–1056.
[3] Felsen, L. B. and Marcuvitz, N. (1973). *Radiation and Scattering of Waves*. Prentice-Hall, Englewood Cliffs, NJ.
[4] Collin, R. E. (1960). *Field Theory of Waves*. McGraw-Hill, New York.

7
INFINITELY THIN STRIP TRANSMISSION LINE STRUCTURES

7.1 Introduction

Single and coupled strip transmission line structures of which the microstrip is the best-known example are widely used in circuit design. A large body of literature has been devoted to the quasi-static and full-wave analysis of the circuit parameters of these structures. For an overview of the different quasi-static techniques used for microstrip design and the presentation of the most advanced techniques we refer the reader to [1] and [2].

The full-wave techniques can be divided into two approaches. The common starting point in both approaches is the integral equation governing the behaviour of the structure. This integral equation follows from the application of the boundary condition of the electric field on the strip surfaces. The electric field is derived from the current flowing on the strip. The relation between the field and the current is established by a suitable Green's dyadic.

A first approach is the widely used spectral domain technique. In this technique the integral equation is transformed to the spectral domain and solved with the method of Galerkin. The current is expanded into a finite number of basis functions and those functions are also used to test the integral equation. In this way, the integral equation is transformed into a linear set of equations. A solution of this set of equations yields the propagation constant of a mode and the associated current. All the other physical quantities can be derived from the current and the propagation constant. References [3]–[12] give an overview of the improvements that the spectral domain technique has undergone since it was first introduced for the modelling of microstrips.

This chapter is devoted to a second approach which was recently discussed by two of the present authors in [13]–[15]. The method presented parallels their published work. In this approach the integral equation is solved in the space domain. The kernel of the integral equation is an appropriate spatial Green's dyadic found as the inverse Fourier transform of a spectral Green's dyadic. The integral equation is solved with the method of moments. To that end the current is expanded into basis functions. In our work we have used a very flexible current modelling based on subsectional basis functions which also takes the edge behaviour of the current into account. Unlike the spectral techniques, our spatial technique does not test the integral equation with the

138   7 THIN STRIP TRANSMISSION LINE STRUCTURES

basis functions. Instead, Dirac pulses are used which makes our method a point matching or collocation method. The combination of a more advanced method of moments, using test functions other than Dirac pulses, and our spatial approach is also possible. However, we have opted for the simple, point matching method. This choice is justified as we use very accurate current basis functions.

This chapter is organized in the following way. After a description of the geometry of the structures under study we construct the integral equation for the current of a mode propagated by such structures. Next, we present the main characteristics of the method of moments solution technique used for the integral equation. Then we proceed with the calculation of the spectral and spatial Green's dyadic. Once we have found the spatial Green's dyadic we move on to the calculation of the contribution of each basis function to the linear set of equations to which the integral equation is transformed by applying the method of moments. We finish this chapter by giving a few case studies which prove the accuracy and the capabilities of the software based on the technique presented in this chapter.

7.2 Geometry of the strip structures

The cross-sectional geometry of a multilayered strip structure is shown in Fig. 7.1. A general structure consists of L layers. The parameters of these layers together with the coordinate system are described in the previous chapter. The metallic strips involved are assumed to be perfectly conducting

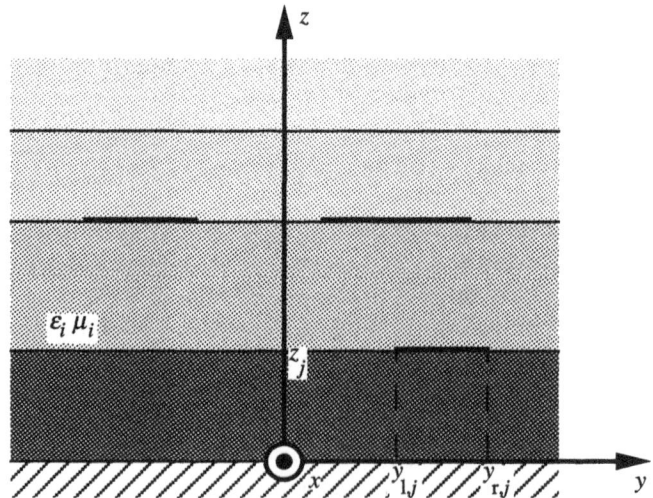

Fig. 7.1. Cross-sectional geometry of a strip structure.

7.3 CONSTRUCTION OF THE INTEGRAL EQUATION

and infinitely thin. The number of strips is denoted by S. The strips are either coplanar, that is, located in the same z-plane, or non-coplanar. The edges of each strip j ($j = 1, \ldots, S$) have y-coordinates denoted by $y_{l,j}$ and $y_{r,j}$ and the transversal position of a strip j is given by the z-coordinate z_j.

7.3 Construction of the integral equation

The integral equation which governs the behaviour of the strip structure follows from the application of the electric boundary condition on each strip. This boundary condition requires the tangential electric field to be zero. This field is given by a convolution integral of the strip current and an appropriate electric Green's dyadic. In this section the integral equation is derived.

For the calculation of the electromagnetic field in sinusoidal regime, excited by a mode of the strip structure, we start from the surface current densities $\mathbf{j}_j = j_{x,j}\mathbf{1}_x + j_{y,j}\mathbf{1}_y$ on each strip j:

$$\begin{aligned} j_{x,j}(x, y, t) &= J_{x,j}(y) \, e^{j(\omega t - \beta x)} \\ j_{y,j}(x, y, t) &= J_{y,j}(y) \, e^{j(\omega t - \beta x)}. \end{aligned} \qquad (7.1)$$

The expression (7.1) for the current $\mathbf{j}_j$ is defined on the surface of strip j. The current $\mathbf{j}_j$ has no z-component as we assumed infinitely thin strips.

The electromagnetic field generated everywhere in the multilayered structure by the current on the strips can be found with the help of suitable Green's dyadics. For example, for the electric field we have

$$\mathbf{e}(x, y, z, t) = \sum_{j=1}^{S} \int_{y_{l,j}}^{y_{r,j}} \mathbf{G}_e(y, z, y', z' = z_j; \beta) \mathbf{J}_j(y') \, e^{j(\omega t - \beta x)} \, dy'. \qquad (7.2)$$

The coordinates (x, y, z) belong to an observation point. The integration variables y' and $z' = z_j$ are the coordinates of an excitation point on strip j. The kernel in the integral representation for the electric field is the Green's dyadic which gives the electric field at an arbitrary point of the structure excited by a line current with the same longitudinal dependence $e^{-j\beta x}$ as the mode under study. The Green's dyadic depends on the coordinates of the observation and excitation point and also on the propagation constant β of the mode. As we can see from (7.2) the electric field takes the general form

$$\mathbf{e}(x, y, z, t) = \mathbf{E}(y, z) \, e^{j(\omega t - \beta x)}. \qquad (7.3)$$

In the following we will suppress the time dependence $e^{j\omega t}$ and the

7 THIN STRIP TRANSMISSION LINE STRUCTURES

longitudinal dependence $e^{-j\beta x}$. The currents $\mathbf{J}_j(y)$ generate an electromagnetic field $[\mathbf{E}(y, z), \mathbf{H}(y, z)]$. Hence for the electric field $\mathbf{E}(y, z)$ we have the following integral representation:

$$\mathbf{E}(y, z) = \sum_{j=1}^{S} \int_{y_{l,j}}^{y_{r,j}} \mathbf{G}_e(y, z, y', z_j; \beta) \mathbf{J}_j(y') \, dy'. \tag{7.4}$$

The Green's dyadic $\mathbf{G}_e$ is the solution to the Maxwell equations and its boundary conditions for a line current in the multilayered medium of the structure but without the strips. Therefore the electric field $\mathbf{E}(y, z)$ is also a solution to the Maxwell equations and fulfils all the boundary conditions at the layer interfaces and the ground planes except the boundary conditions at the strips themselves as they are not incorporated in the Green's dyadic. A similar result holds for the magnetic field which is also given by an integral representation such as (7.4) but with a different Green's dyadic. However, this dyadic also fulfils the magnetic boundary condition at the line source, that is, the dyadic will exhibit a jump which corresponds to the complex amplitude of the current. So, the integral representation for the magnetic field also incorporates the boundary conditions at the strips.

In order to find the unknown currents $\mathbf{J}_j$ and the propagation constant β of a mode we impose the remaining boundary condition. We require the tangential electric field on the perfectly conducting strips to be zero. This leads to the integral equation for the mode:

$$\lim_{z \to z_o} \sum_{j=1}^{S} \int_{y_{l,j}}^{y_{r,j}} \mathbf{G}(y, z, y', z_j; \beta) \mathbf{J}_j(y') \, dy' = \mathbf{0}. \tag{7.5}$$

This can be rewritten in terms of two coupled scalar equations:

$$\lim_{z \to z_o} \sum_{j=1}^{S} \int_{y_{l,j}}^{y_{r,j}} (G_{xx} J_{j,x} + G_{xy} J_{j,y}) \, dy' = 0$$

$$\lim_{z \to z_o} \sum_{j=1}^{S} \int_{y_{l,j}}^{y_{r,j}} (G_{yx} J_{j,x} + G_{yy} J_{j,y}) \, dy' = 0. \tag{7.6}$$

The observation point (y, z) is now restricted to the strips. $\mathbf{G}$ represents this part of $\mathbf{G}_e$ which yields the tangential x- and y-components of $\mathbf{E}$; it is therefore a 2×2 matrix. The integral equation follows from (7.4) if one lets the observation point approach the strip. In (7.5) the observation point approaches the strip along a path perpendicular to and above the strip. However, the value of the left-hand side of (7.5) does not depend on the choice of that path. The final value of the z-coordinate of the observation point after the limit operation is denoted by z_o in (7.5). It is essential to retain that limit. We will return to this later on in the chapter.

7.4 Method of moments solution technique

The integral equation (7.5) is solved with the help of the method of moments. This method has stood the scientific test of time and has proven to be very efficient and accurate for the problem with which we are dealing here. In this method the x- and y-component of the unknown currents $\mathbf{J}_j$ are expanded into a finite sum of known basis functions with unknown complex amplitudes. These amplitudes replace the current as the unknowns of the problem (beside the propagation constant). In order to find the amplitudes we test both the x- and y-component of the integral equation on a certain strip with a number of test functions equal to half the number of basis functions used in the expansion of the current components on that strip. These two steps, that is, the expansion of the unknown current in basis functions and the testing of the integral equation, transform the original integral equation into a linear set of equations. This set forms a discrete eigenvalue problem in which the propagation constant β and the vector with the amplitudes of the basis functions are the eigenvalue and the eigenvector, respectively.

We will apply, but not explain, the method of moments in this book. The reader who is not familiar with this method is referred to the standard book of Harrington on the method of moments [16]. In the next three sections we will discuss the two above-mentioned steps in the method of moments and the solution of the discrete eigenvalue problem in more detail.

7.4.1 Current basis functions

For the expansion of the unknown current components into a finite set of independent basis functions we divide each strip j into N_j equal intervals of width Δ_j. The number of intervals N_j may differ from strip to strip. We approximate the current components by a continuous and piecewice linear function in the intervals 2 to $N_j - 1$. In the outermost intervals 1 and N_j another approximation is used which models the behaviour of the current near the edge of a metal plate. This behaviour has been analysed by Meixner and imposes the following general form for the longitudinal current [17]:

$$J_{x,j} = A_j \tau^{-\frac{1}{2}} + B_j \tau^{\frac{1}{2}} + C_j \tau^{\frac{3}{2}} + O(\tau^{\frac{5}{2}}). \tag{7.7}$$

where τ represents the distance to the edge. This component of the current becomes infinite at the edge. The transversal component remains finite and becomes zero at the edge:

$$J_{y,j} = B'_j \tau^{\frac{1}{2}} + C'_j \tau^{\frac{3}{2}} + O(\tau^{\frac{5}{2}}). \tag{7.8}$$

In our approximation of the current we restrict the series (7.7) to the first three terms and the series (7.8) to the first two terms.

In order to ensure the continuity of the modelled current along the strip the current is expanded into triangular functions $t_{tr}(y)$ in the intervals 2 to $N_j - 1$ and modified triangular functions $t_{1,j}(y)$, $t_{2,j}(y)$, and $t_{3,j}(y)$ for the longitudinal component and $t_{2,j}(y)$ and $t_{3,j}(y)$ for the transversal component in the two outermost pairs of intervals. The triangular functions $t_{tr}(y)$ in the interval (y_L, y_R) on strip j are defined as

$$t_{tr}(y) = \frac{y - y_L}{\Delta_j} \qquad y_L \leq y \leq y_L + \Delta_j = y_R - \Delta_j$$

$$t_{tr}(y) = \frac{y_R - y}{\Delta_j} \qquad y_R - \Delta_j \leq y \leq y_R \qquad (7.9)$$

$$y_R - y_L = 2\Delta_j.$$

The modified triangular basis functions incorporate the Meixner edge condition. They are given by

$$\begin{aligned} t_{1,j}(\tau) &= (\Delta_j/\tau)^{\frac{1}{2}} \\ t_{2,j}(\tau) &= (\tau/\Delta_j)^{\frac{1}{2}} \qquad\qquad 0 \leq \tau \leq \Delta_j \\ t_{3,j}(\tau) &= (\tau/\Delta_j)^{\frac{3}{2}} \end{aligned} \qquad (7.10)$$

$$t_{1,j}(\tau) = t_{2,j}(\tau) = t_{3,j}(\tau) = \frac{(2\Delta_j - \tau)}{\Delta_j} \qquad \Delta_j \leq \tau \leq 2\Delta_j$$

where τ is the distance to either edge. Figure 7.2 displays the different types of basis functions and their superposition for the longitudinal current $J_{x,j}(y)$ at the left side, and for the transversal current $J_{y,j}(y)$ at the right side. The current components $J_{x,j}(y)$ and $J_{y,j}(y)$ on strip j are a superposition of $N_j + 3$ and $N_j + 1$ basis functions, respectively. For lossless structures one can easily show that both components are in quadrature.

The basis functions used in the expansion of the current are defined only on a subsection of the strip and are called subsectional basis functions. In the literature solution techniques can be found that use entire domain basis functions, that is, basis functions defined over the entire strip. Typical examples of this approach can be found in [4] and [10].

7.4.2 Testing of the integral equation

The total number of scalar unknowns introduced in the expansion of the current components on strip j is $2N_j + 4$. To determine these unknowns we will test both components of the integral equation with $N_j + 2$ test functions

7.4 METHOD OF MOMENTS SOLUTION TECHNIQUE

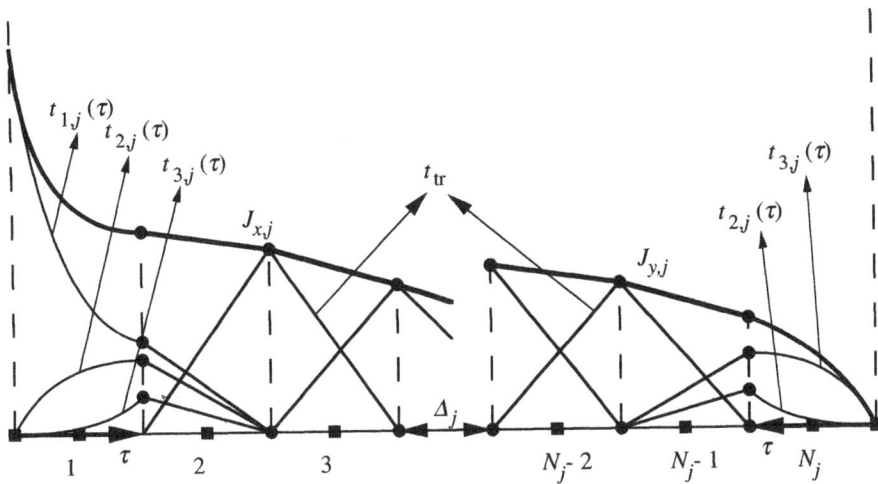

Fig. 7.2. Longitudinal current $J_{x,j}$ and transversal current $J_{y,j}$ as a superposition of elementary basis functions.

on strip j. In our technique we use the Dirac pulse functions to test the integral equation. Therefore we require the integral equation to be satisfied at $N_j + 2$ points on strip j. These sample points are chosen to be in the middle of each interval and at the edges of the strip. This approach is called the point matching or collocation method. The sample points are marked by a square dot on Fig. 7.2.

We note that our choice of test functions differs from the choice made in the spectral domain method. In this method the test and basis functions are the same and the method is therefore called a Galerkin method of moments. However, unlike the Galerkin method of moments, the point matching technique has no variational properties. Because of the very accurate current modelling we will prove in the case studies that we obtain excellent results with the simpler-to-implement point matching technique.

7.4.3 Solution of the discrete eigenvalue problem

In the following the coordinates of each matching point will be denoted by (y, z_o). This distinguishes these matching points from the more general observation point (y, z) in the argument of the Green's dyadic $\mathbf{G}$ in (7.5). In enforcing (7.5) in each matching point, we have to calculate the contribution of the current on each strip to the tangential electric field at that point. In turn, the current on each strip and hence the electric field generated by that current, is constituted by the different contributions coming from each basis function used to discretize that current. In Section 7.4.1 different types of basis functions were introduced. Here they will be denoted by the common

144   7 THIN STRIP TRANSMISSION LINE STRUCTURES

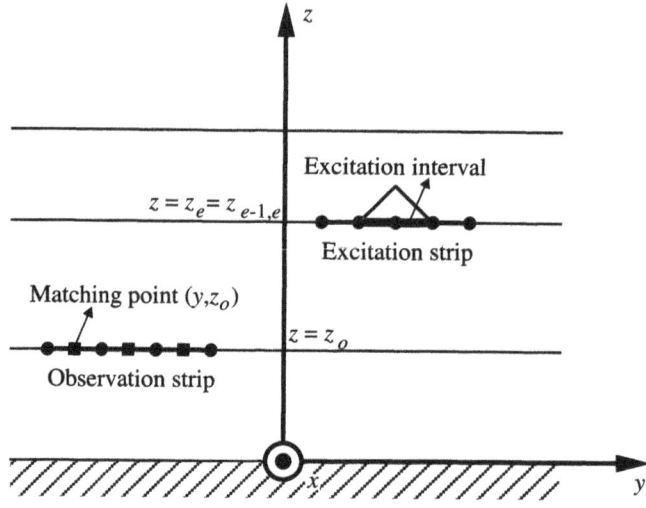

Fig. 7.3. Excitation and observation strip.

symbol $t(y')$ where the argument y' shows that each basis function depends on a running coordinate along one of the strips. Each basis function differs from zero only in a finite y' interval. Hence, the current associated with each basis function will contribute to (7.5) via the integration over a finite y' interval. This interval or in other words this section somewhere on one of the strips will be called the *excitation interval*. The strip on which the matching point is located is called the *observation strip*. The excitation and observation strip are depicted on Fig. 7.3. From the above reasoning it is now clear that in order to obtain (7.5) for a particular matching point we have to sum up the contributions from all excitation intervals for all basis functions and for all strips. The final unknowns of the problem are the complex amplitudes of each basis function. The method of moments described above reduces the original integral equation to a set of linear equations for these complex amplitudes, that is, one equation per matching point. We will not try to represent the whole reasoning given above in complex mathematical expressions but will restrict ourselves to giving a typical integral that arises while discretizing (7.5) into a set of linear equations. This integral is given by (see also Fig. 7.3):

$$\lim_{z \to z_o} A \int_{\text{excitation interval}} G_{ab}(y, z, y', z_e; \beta) t(y') \, dy'. \qquad (7.11)$$

The height at which the excitation strip is located is denoted by z_e and A is the complex amplitude of the basis function $t(y')$ on the excitation strip. (y, z_o) are the fixed coordinates of one particular matching point. (y', z_e) are

7.5 THE SPECTRAL GREEN'S DYADIC

the coordinates of a running point on the excitation interval. Looking at (7.6), one immediately concludes that the four elements of the Green's dyadic **G** are needed. This is reflected in (7.11) by the subindex ab where a and b are either x or y. The integration over y' and the limit operation still have to be performed and the Green's function G_{ab} must be determined. The integration will be performed in the space domain as already mentioned in Section 7.1. The spatial Green's dyadic will be found as the inverse Fourier transform of the spectral dyadic. In Sections 7.5 and 7.6 the calculation of the spectral and spatial Green's dyadic is treated. In Section 7.7 we come back to the spatial integration in (7.11).

7.5 The spectral Green's dyadic

This section makes extensive use of Chapter 6 on the spectral field in a multilayered medium. If we are looking for the elements G_{ab} of **G** we are really looking for the x- and y-components of the electric field everywhere in the layered medium and in particular on the observation strips generated by a unit line current located at a particular point of an excitation strip. As shown in Chapter 6, the calculation of the spectral field excited by such a source can be reduced to solving a TE and TM cascade. As a first step, the excitation of both cascades has to be determined from the boundary conditions for the spectral field at the interface where the line source is located. Next, the excited spectral field can be found by using classical transmission line techniques for the solution of the TE and TM cascade. Finally, from the wave pattern in the TE and TM cascade the Green's dyadic associated with the line current can be derived. These three steps will be elaborated in the next three sections.

7.5.1 The excitation of the TE and TM cascade

In the spectral domain the boundary conditions at the interface where the line source is located can easily be derived from their spatial counterpart. In the $(', '', z)$ coordinate system we have:

$$\begin{aligned} E'_e - E'_{e+1} &= 0 \\ E''_e - E''_{e+1} &= 0 \\ H'_e - H'_{e+1} &= -J''(k_y) \\ H''_e - H''_{e+1} &= J'(k_y). \end{aligned} \quad (7.12)$$

$\mathbf{J}(k_y) = J_x(k_y)\mathbf{1}_x + J_y(k_y)\mathbf{1}_y$ is the Fourier transform of a current located in the plane $z = z_{e,e+1}$, that is, at the height of one particular excitation strip.

146   7 THIN STRIP TRANSMISSION LINE STRUCTURES

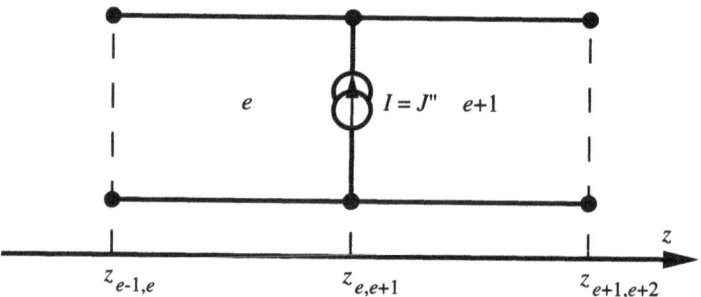

Fig. 7.4. Excitation of the TE cascade.

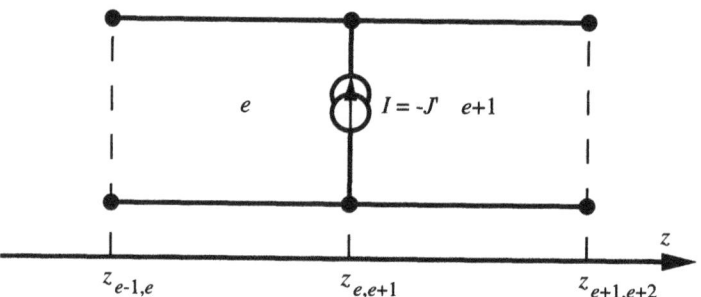

Fig. 7.5. Excitation of the TM cascade.

It is assumed that this strip is located between the layers e and $e + 1$. In Chapter 6 the notation $z_{e,e+1}$ was introduced to indicate the z-coordinate of the interface between layer e and layer $e + 1$. In the present chapter, and because subsequent confusion is impossible, we simply replace $z_{e,e+1}$ by z_e. The excitations in the TE and TM cascade follow directly from (7.12) and are sketched in Figs 7.4 and 7.5, respectively. The point at which the source is located will be called the excitation point because such a point is always located on an excitation strip.

7.5.2 *Solution of the TE and TM cascade*

Both cascades can be solved in a similar way as we have shown in Chapter 6. We will therefore use their generic notation and solve the generic cascade. In Chapter 6 we constructed a mechanism to find the wave coefficients in all the layers. Here we start from the boundary conditions at the interface of the strip. For convenience we repeat the boundary condition equations (6.46):

$$\left(\prod_{k=L}^{e+1} \mathbf{N}_k^d\right) \mathbf{K}_L \mathbf{R}_L - \left(\prod_{k=1}^{e} \mathbf{N}_k^u\right) \mathbf{K}_1 \mathbf{R}_1 = \begin{bmatrix} 0 \\ I \end{bmatrix}. \tag{7.13}$$

7.5 THE SPECTRAL GREEN'S DYADIC

The excitation of the cascade is a current source I located at $z = z_e$ (see Figs 7.4 and 7.5). There is no voltage excitation. From the solution of the two scalar linear equations (7.13) follow the coefficients R_1 and R_L and the spectral field everywhere in the cascade using the transformation formulas (6.47) and (6.48). For an observation point z located in a layer i below the strip we have

$$\mathbf{C}_i(z) = e^{-\Gamma_i(z_{i,i+1}-z)}\left(\prod_{k=e}^{i+1} e^{-\Gamma_k d_k}\right) \mathbf{N}_i^u(z) \left(\prod_{k=1}^{i-1} \mathbf{N}_k^u\right) \mathbf{K}_1 R_1 \qquad (7.14)$$

and

$$\mathbf{C}_i(z) = e^{-\Gamma_i(z-z_{i-1,i})}\left(\prod_{k=e+1}^{i-1} e^{-\Gamma_k d_k}\right) \mathbf{N}_i^d(z) \left(\prod_{k=L}^{i+1} \mathbf{N}_k^d\right) \mathbf{K}_L R_L \qquad (7.15)$$

for an observation point z located in a layer i above the strip. The equations (7.14) and (7.15) describe the spectral field in the cascade. This field can be generated numerically in a fast and efficient way.

7.5.3 Construction of the Green's dyadic

For the calculation of the integrals (7.11) we need only the Green's dyadic associated with the x- and y-components of the electric field. Those two components can be derived from E'_i and E''_i which follow directly from the solution of the TE and TM cascades. Remember that the subscript i refers to the layer in which the fields are calculated. In order to construct the Green's dyadic we write E'_i and E''_i in terms of the current excitations J' and J'':

$$\begin{aligned}E'_i(z) &= G_{\text{TM}}(z)J' \\ E''_i(z) &= G_{\text{TE}}(z)J''.\end{aligned} \qquad (7.16)$$

The impedance $G_{\text{TM}}(z)$ is the voltage propagating in the TM cascade for a current excitation of -1 ampere at the interface $z = z_e$. The impedance $G_{\text{TE}}(z)$ is the voltage propagating in the TE cascade for a current of 1 ampere at the interface $z = z_e$. Both impedances follow from (7.14) and (7.15). These impedances are in fact scalar Green's functions in the spectral domain relating ' and " components of the source current to the corresponding components of the electric field everywhere in the layered medium.

The spectral Green's dyadic, $\mathbf{G}$, of interest describes the relation between the x- and y-components of the spectral electric field and the spectral current:

$$\begin{bmatrix} E_x \\ E_y \end{bmatrix} = \begin{bmatrix} G_{xx} & G_{xy} \\ G_{yx} & G_{yy} \end{bmatrix} \begin{bmatrix} J_x \\ J_y \end{bmatrix}. \qquad (7.17)$$

148   7 THIN STRIP TRANSMISSION LINE STRUCTURES

The four components of **G** in (7.17) can easily be expressed in terms of the two impedances G_{TE} and G_{TM} using the relation between the components of an arbitrary vector in the cartesian and in the $(',\,'',\,z)$ coordinate systems. Simple mathematics results in the following form for the four components of **G**:

$$G_{xx} = \frac{G_{TM}(z)\beta^2 + G_{TE}(z)k_y^2}{k^2}$$

$$G_{xy} = G_{yx} = \frac{[G_{TM}(z) - G_{TE}(z)]k_y\beta}{k^2} \qquad (7.18)$$

$$G_{yy} = \frac{G_{TM}(z)k_y^2 + G_{TE}(z)\beta^2}{k^2}$$

with

$$k^2 = \beta^2 + k_y^2. \qquad (7.19)$$

To conclude this section on the spectral Green's dyadic we discuss some important properties of the impedances $G_{TE}(z)$ and $G_{TM}(z)$.

These impedances are even functions of k_y. This follows from the fact that the propagation constant and the characteristic impedance of each layer in a cascade depend only on the Γ function of the layers which in turn is a function only of k_y^2.

As we mentioned in Chapter 6, the impedances are also even function in Γ at least for bounded layers, that is, for all the layers except the semi-infinite outermost ones [18]. As a consequence, only these semi-infinite layers give rise to branch cuts in the complex k_y plane if we consider the value of the impedance functions $G_{TE}(z)$ and $G_{TM}(z)$ for complex k_y values.

In each point of a cascade, that is, at each height in the layered medium, $G_{TE}(z)$ and $G_{TM}(z)$ can be written as an infinite series of travelling waves. We distinguish a direct wave and indirect waves. The direct wave propagates straight from the excitation to the observation point. The distance travelled by the wave corresponds to the absolute difference in z-coordinates of the excitation and the observation point. In each layer the direct wave is attenuated by a factor $e^{-\Gamma d}$, with Γ and d the spectral propagation constant and the thickness of the layer, respectively. At each interface the wave is also attenuated by the transmission coefficient of that interface. For large k_y values the direct wave exhibits an exponential decay for observation points located in a z-plane different from the z-plane of the excitation. In the case where the observation point and the excitation point are located in the same z-plane this exponential decay is not present because the distance travelled by the wave becomes zero. The indirect waves are the waves which have reflected at least once. For large k_y values they always exhibit an exponential

7.6 The spatial Green's dyadic

7.6.1 *The inverse Fourier transformation*

The spatial Green's dyadic follows from the inverse Fourier transformation of the spectral dyadic calculated in the previous section. This means that we have to calculate the inverse Fourier transforms of G_{xx}, G_{xy}, and G_{yy} in (7.18). This inverse Fourier transform is defined by equation (6.12). From the translational invariance of the layered medium in the y-direction it follows that the spatial Green's function **G** will depend only on the distance in the y-direction between excitation and observation point. This means that G_{ab} in (7.11) depends only upon $y - y'$. Taking the above and the definition of the inverse Fourier transform into account, we can deduce from (7.18) that the calculation of the inverse Fourier transformation requires the calculation of two sets of integrals:

$$I_{TE}^p(\delta, z; \beta) = \int_{-\infty}^{+\infty} \frac{G_{TE}(z) k_y^p}{k^2} e^{-jk_y \delta} dk_y \qquad (7.20)$$

and

$$I_{TM}^p(\delta, z; \beta) = \int_{-\infty}^{+\infty} \frac{G_{TM}(z) k_y^p}{k^2} e^{-jk_y \delta} dk_y. \qquad (7.21)$$

The notation $\delta = y - y'$ stands for the distance in the y-direction between the observation point and the excitation point. The first set of integrals (7.20) is associated with the impedance G_{TE} which characterizes the TE cascade and the integrals are therefore called the TE integrals. The power p of k_y takes the values 0, 1, and 2. Analogously, the set of integrals in (7.21) are called the TM integrals.

The elements of the spatial Green's dyadic (7.18) can be expressed in terms of the TE and TM integrals:

$$\begin{aligned} G_{xx}(\delta, z; \beta) &= \beta^2 I_{TM}^0(\delta, z; \beta) + I_{TE}^2(\delta, z; \beta) \\ G_{xy}(\delta, z; \beta) &= G_{yx}(\delta, z; \beta) = \beta[I_{TM}^1(\delta, z; \beta) - I_{TE}^1(\delta, z; \beta)] \\ G_{yy}(\delta, z; \beta) &= I_{TM}^2(\delta, z; \beta) + \beta^2 I_{TE}^0(\delta, z; \beta). \end{aligned} \qquad (7.22)$$

In order to perform the inverse Fourier integrations in (7.20) and (7.21) the

150   7 THIN STRIP TRANSMISSION LINE STRUCTURES

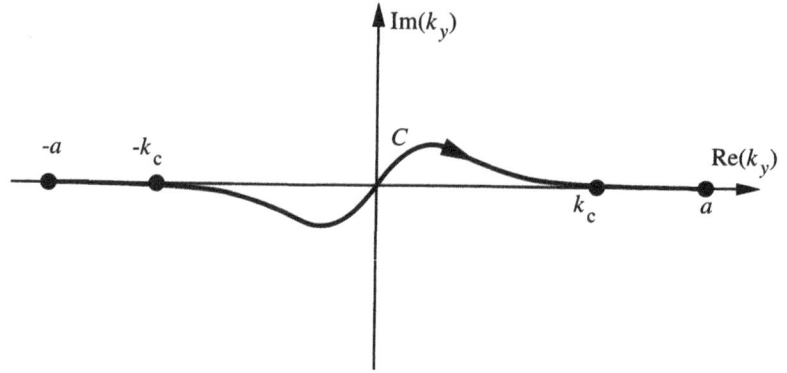

Fig. 7.6. Complex integration path for the TE and TM integrals.

real variable k_y is extended to a complex variable, as discussed in Chapter 6. The integration path C of the TE and TM integrals in the complex k_y plane is shown on Fig. 7.6. The integration path is divided into three parts: two infinite integration intervals, $]-\infty, -k_c]$ and $[k_c, +\infty[$ on the real axis, and one finite integration interval $[-k_c, k_c]$ in the complex plane. The value of k_c is chosen in such a way that all the Γ functions can be approximated by $|k_y|$ while $\beta^2 + k_y^2$ can be replaced by k_y^2 in the two infinite integration intervals on the real axis. In the next section we discuss the evaluation of the TE and TM integrals. We will focus the discussion on the evaluation of the TE and TM integrals for coinciding observation and excitation plane, that is, for $z_e = z_o$. This is by far the most difficult situation from a numerical point of view.

7.6.2 *The evaluation of the TE and TM integrals*

The components of the spatial Green's dyadic are used in the calculation of the contributions (7.11) from each excitation interval. Therefore we can restrict the evaluation of the spatial Green's dyadic to matching points (y, z_o). However, in certain cases special care should be taken in handling the limit operation in (7.11).

We first turn our attention to the case where the observation strip and the excitation strip are located at different heights, that is, $z_e \neq z_o$. In that case the direct wave and the indirect waves in both impedances G_{TE} and G_{TM} exhibit an exponential decay for large absolute k_y values as already explained at the end of Section 7.5.3. As a consequence, the TE and TM integrals have an integrand which decays exponentially as a function of k_y for large absolute k_y values. This, along with the regularity of the integrand, guarantees the convergence of the TE and TM integrals, in particular for the infinite

7.6 THE SPATIAL GREEN'S DYADIC

integration intervals on the real axis. The behaviour of the integrands and the associated convergence of the integrals also implies that the limit operation in (7.11), where a general observation point approaches the strip to become a matching point on that strip, can be interchanged with the space domain integration in (7.11) and with the inverse Fourier integration in (7.20) or (7.21). In the final numerical approach adopted by us, the infinite integration path of Fig. 7.6 is truncated to a path $[-a, a]$ of finite length were a is chosen in such a way that the contribution from $]-\infty, -a]$ and from $[a, +\infty[$ can be neglected. The points $-a$ and a are located on the real axis, $a > k_c$; see Fig. 7.6. The integration along the path $[-a, a]$, partly on the real axis, partly in the complex plane, can be performed with Gaussian quadrature formulas. Details are provided in [13]. More sophisticated techniques are also available in the literature; see for example [19].

We now turn our attention to the case where the excitation strip and observation strip are at the same height, that is, $z_e = z_o$. For such a case the calculation of the TE and the TM integrals is more complicated. Again, both G_{TE} and G_{TM} consist of a direct wave and an infinite set of indirect waves. All the indirect waves exhibit an exponential decay for large enough $|k_y|$ values. However, the direct wave does not exhibit an exponential decay. As we will prove in the next section, the integrand of the TE and TM integrals for large $|k_y|$ values and a z position just above the observation strip, consists of a sum of contributions of the form

$$k_y^m \, e^{-jk_y\delta} \, e^{-|k_y|(z-z_e)} \, \text{sign}(k_y) \tag{7.23}$$

with $m = p - 3 = -3, -2, -1$ for the TE integrals and $m = p - 1 = -1, 0, 1$ for the TM integrals. In (7.23) we have that $\text{sign}(k_y) = 1$ for $k_y > 0$ and $\text{sign}(k_y) = -1$ for $k_y < 0$. We now still have to replace z in (7.23) by z_o in order to perform the limit operation prescribed in (7.11). As far as the convergence of the $]-\infty, -k_c]$ and $[k_c, +\infty[$ contributions to the TE and TM integrals is concerned, we can draw the following two conclusions from (7.23).

First, for different y-coordinates of the excitation and matching point but coinciding z-coordinates, that is, $\delta \neq 0$ and $z = z_o = z_e$ in (7.23), the TM integrals for p equal to 1 and 2 (that is, $m = 0$ or 1) are divergent while all the other TE and TM integrals are convergent. Thus, for the convergent integrals the limit operator in (7.11) can again be allowed to operate directly on the integrand and hence (7.23) simply becomes $k_y^m \, e^{-jk_y\delta} \, \text{sign}(k_y)$. For the divergent integrals, however, we will show in the next section that the limit operator in (7.11) can still be interchanged with the spatial integration but cannot be interchanged with the inverse Fourier integration in (7.20) or (7.21).

Second, for coinciding matching and excitation point, that is, $\delta = 0$ and $z_o = z_e$, the TE integrals for p equal to 0 and 1 (that is, $m = -3$ or -2) are convergent while all the other TE and TM integrals are divergent. For the convergent integrals (7.23) becomes $k_y^m \operatorname{sign}(k_y)$ and z can be replaced by z_o in (7.20) and (7.21). For the divergent integrals we will show that the limit operator in (7.11) cannot be interchanged with the spatial integration, that is, that we have to calculate the integral first with $z \neq z_o$ and only replace z by z_o in the result of that integration.

We will now discuss how the inverse Fourier transformation is performed when the excitation and observation strip are at the same height, that is, for $z_o = z_e$.

We write G_{TE} and G_{TM} as the sum of a direct wave and a remainder part. The latter part represents all indirect waves. The contribution of the remainder part to the TE and TM integrals can be calculated in exactly the same way as the TE and TM integrals for non-coinciding excitation and observation planes, that is, by using Gaussian quadrature formulas. This is a consequence of the exponential decay which is present in those remainder parts for large absolute k_y values. For these remainder parts the limit operation can always operate directly on the integrands of (7.20) and (7.21).

For the calculation of the contribution of the direct wave we distinguish two contributions. The first one comes from the integration along the path in the complex plane (see Fig. 7.6) and is therefore a finite integral. As the integrand is a regular function the limit operation in (7.11) can again be interchanged with the integration in (7.11) and the integrations in (7.20) and (7.21). Consequently, the integral can be calculated numerically without any difficulty. The second contribution comes from the two integrations on the real axis. For these integrals the limit operation cannot operate directly on the integrands as those integrands take the asymptotic form (7.23) which results for certain cases in divergent integrals as discussed above. In the following three sections we will calculate the direct wave at an observation point just above the excitation strip, derive the contribution of the direct wave to the TE and TM cascade, and determine the singular behaviour of the Green's dyadic. Only the main steps will be given.

7.6.2.1 Calculation of the direct wave for coinciding observation and excitation plane The direct wave follows from the solution of the TE and TM cascade associated with the double layered structure shown in Fig. 7.7. The two layers are semi-infinite in the z-direction and the layer characteristics correspond to the characteristics of the layers e and $e + 1$ of the original TE and TM cascades. Hence, we now have a very simple layered medium that can easily be handled analytically (in the spectral domain) with the techniques discussed in Chapter 6. As the direct wave is defined in such a way that it is in fact the wave which encounters no reflections when leaving the source, it is immediately clear that this direct wave can be calculated from the simple structure depicted in Fig. 7.7. We restrict ourselves to giving the final results,

7.6 THE SPATIAL GREEN'S DYADIC

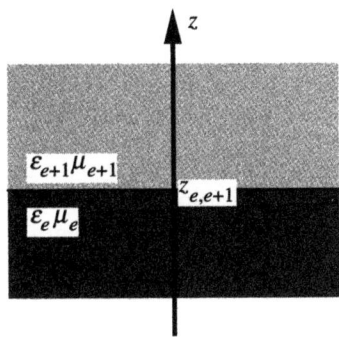

Fig. 7.7. Double layered medium used for the calculation of the direct wave.

that is, the spectral Green's functions as defined in (7.16) but for the configuration of Fig. 7.7. For the TM direct wave we find the following impedance G_{TM}^{d} above the excitation, that is, in the layer $e + 1$:

$$G_{\text{TM}}^{\text{d}} = \frac{\Gamma_e \Gamma_{e+1}}{j\omega(\varepsilon_{e+1}\Gamma_e + \varepsilon_e \Gamma_{e+1})} e^{-\Gamma_{e+1}(z-z_e)}. \qquad (7.24)$$

The superscript 'd' refers to the direct wave. For the TE cascade we find a similar solution for the impedance G_{TE}^{d} above the excitation:

$$G_{\text{TE}}^{\text{d}} = -\frac{j\omega\mu_e\mu_{e+1}}{\mu_e\Gamma_{e+1} + \mu_{e+1}\Gamma_e} e^{-\Gamma_{e+1}(z-z_e)}. \qquad (7.25)$$

The reader can easily derive these formulas from the general wave representation in a layer of the TE and TM cascade and the boundary conditions at the current excitation. We have restricted the result to the situation where the observation point is located above the excitation, that is, to $z > z_e$. The reason for this is that the limit $z \to z_o$ in (7.5) and (7.11) is taken along a path parallel to the y-axis and starting from a z value greater than z_o. As already mentioned in Section 7.3, any other path would yield the same result.

7.6.2.2 Contribution of the direct wave to the TE and TM integrals If we now replace G_{TE} and G_{TM} in (7.20) and (7.21) with the corresponding expressions for the direct wave (7.24) and (7.25) it is clear that these integrals are only convergent in all cases provided z is still different from z_e. As discussed above, the problems only arise from the asymptotic parts of the integrals, that is, those parts where $|k_y| > k_c$. We will now extract these asymptotic parts, calculate them analytically for an arbitrary z value, and then also perform the limit operation $z \to z_o$ analytically. To find the

154   7 THIN STRIP TRANSMISSION LINE STRUCTURES

asymptotic part we replace k^2 by k_y^2 and Γ_i, with $i = e$ or $i = e + 1$, by

$$\Gamma_i \to k_y\left(1 + \frac{\gamma_i^2}{2k_y^2}\right) \tag{7.26}$$

with $\gamma_i^2 = \omega^2 \varepsilon_i \mu_i - \beta^2$. From (7.20)–(7.26) we can see that the calculation of the asymptotic parts of the integrations on the real axis can be reduced to the calculation of the sum of the following integrals:

$$\int_{k_c}^{+\infty} k_y^m e^{-k_y \Delta z - jk_y \delta} dk_y + (-1)^{m+1} \int_{k_c}^{+\infty} k_y^m e^{-k_y \Delta z + jk_y \delta} dk_y \tag{7.27}$$

where $\Delta z = z - z_e$ represents the distance between the matching and excitation point in the z-direction and where m takes the values $m = -5, -4, -3, -2, -1, 0, 1$. Let's take a closer look at the integrals in the first term of (7.27). The conclusions we will draw are of course also valid for the integrals in the second term. For $m = -5, -4, -3, -2$ these integrals in (7.27) are convergent, independent of the value of δ and Δz. The integrals for $m = -1, 0, 1$ which can become divergent, can easily be calculated analytically. The results are

$$\int_{k_c}^{+\infty} k_y^{-1} e^{-k_y \Delta z - jk_y \delta} dk_y = \text{Ei}[k_c(\Delta z + j\delta)]$$

$$\int_{k_c}^{+\infty} k_y^0 e^{-k_y \Delta z - jk_y \delta} dk_y = \frac{e^{-k_c(\Delta z + j\delta)}}{\Delta z + j\delta} \tag{7.28}$$

$$\int_{k_c}^{+\infty} k_y^1 e^{-k_y \Delta z - jk_y \delta} dk_y = \left[\frac{k_c}{\Delta z + j\delta} + \frac{1}{(\Delta z + j\delta)^2}\right] e^{-k_c(\Delta z + j\delta)}.$$

Ei(z) represents the exponential integral [20]:

$$\text{Ei}(z) = \int_z^{+\infty} \frac{e^{-x}}{x} dx = -\gamma - \ln z - \sum_{n=1}^{+\infty} \frac{(-1)^n z^n}{nn!} \tag{7.29}$$

with $\gamma = 0.5772156649\ldots$ the Euler constant. We note that the exponential integral (7.29) has a logarithmic singularity for $z = 0$. For $m = 1$ we have δ^{-2} and δ^{-1} singularities in the integral (7.28) for coinciding observation and excitation planes. This $m = 1$ integral will be needed in the calculation of the TM integral (7.21) for $p = 2$. For $m = 0$ we have only a δ^{-1} singularity in the integral (7.28). This term is found in the TM integral for $p = 1$. For $m = 0$ we have a logarithmic singularity which occurs in the TM integral for $p = 0, 2$ and in the TE integral for $p = 2$.

The contribution of the remainder terms, that is, the complete integrand for the complex part of the path in Fig. 7.6 (between $-k_c$ and k_c) and the difference between the complete integrand and the asymptotic part for the real part of the path in Fig. 7.6 (that is, $k_y < -k_c$ and $k_y > k_c$) is found numerically with the limit operation performed before the integration.

We can conclude that the TE and TM integrals, and thus also the components of the Green's dyadic, become infinite only for coinciding matching and excitation points. Even when the matching point is located in the same plane as the excitation but does not coincide with the excitation the Green's dyadic remains finite. In that case, this finite result could only be obtained by performing the limit operation after the inverse Fourier integration.

7.6.2.3 Singular behaviour of the Green's dyadic From the singular behaviour of the TE and TM integrals and the expressions (7.22) for the components of **G** as a function of the TE and TM integrals we can derive the singular behaviour for the components of the Green's dyadic for coinciding matching and excitation points. The singular terms in the components of the Green's dyadic are shown in Table 7.1.

Table 7.1 Singular terms in the Green's dyadic for coinciding matching and excitation points

G_{xx}: $\ln|\delta|$
G_{xy}, G_{yx}: $\dfrac{1}{\delta}$
G_{yy}: $\dfrac{1}{\delta^2}, \dfrac{1}{\delta}, \ln|\delta|$

7.7 Calculation of the contribution of each basis function to the integral equation

The final step of our calculation technique is the evaluation of the contribution of each basis function (7.11) to the integral equation (7.5) or (7.6) using the components of the spatial Green's dyadic calculated in the previous section. In our discussion we distinguish selfpatch and non-selfpatch contributions. Non-selfpatch contributions are those contributions for which the matching point is not located in the excitation interval, that is, either $z_o \neq z_e$ or $z_o = z_e$ but $(y - y')$ is always different from zero in the integral (7.11). For selfpatch contributions the matching point is located in the excitation interval. We will briefly discuss the calculation technique for the non-selfpatch and the selfpatch contributions.

156   7 THIN STRIP TRANSMISSION LINE STRUCTURES

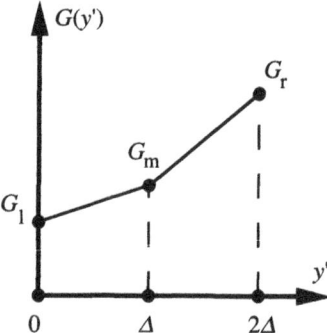

Fig. 7.8. Linear approximation of a component of the Green's dyadic.

7.7.1 *Non-selfpatch contributions*

For the non-selfpatch contributions the components of the Green's dyadic remain finite. For each contribution the component of the Green's dyadic which occurs as the kernel in the integral representation (7.11) is approximated in a way depending on the relative location of the matching point.

In the case where the matching point is not located in the same plane as the excitation interval ($z_o \neq z_e$) or located in the same plane ($z_o = z_e$) but far enough from the excitation interval, typically a few strip intervals, the components of the Green's dyadic are approximated by a piecewise linear function. In Fig. 7.8 this approximation is illustrated, the excitation interval is $[0, 2\Delta]$, and three sample points are used to approximate the Green's dyadic component, two at the edges and one in the middle of the interval.

In the case that the matching point is located in the plane of the excitation interval ($z_o = z_e$) and close enough to the excitation interval (so-called neighbour coupling) a similar linear approximation is used as in the previous case but when a singular term δ^{-1} or δ^{-2} occurs in the Green's dyadic component it is explicitly taken into account. These singular terms do not become infinite for the non-selfpatch contributions but a linear approximation of these terms is not accurate enough to yield stable and accurate mode current profiles. Experience has shown that unlike the eigencurrent, the propagation constant is not affected by a less accurate linear approximation of the singular terms. This is a consequence of the variational character of the propagation constant. The weak logarithmic singularity in the first expression of (7.28) (see also (7.29)) can be approximated by a linear function for the non-selfpatch contributions.

With the approximations introduced above all the non-selfpatch contributions can be calculated analytically or in a numerically efficient and accurate way. Because the elementary intervals on the same strip have the same length

7.7 CONTRIBUTIONS OF BASIS FUNCTIONS

we can reduce the number of Green's dyadic evaluations by exploiting the fact that the integrand in (7.11) depends only upon $(y - y')$.

7.7.2 Selfpatch contributions

Unlike the components for the non-selfpatch contributions, the components of the Green's dyadic become singular for the selfpatch contributions as the matching and excitation points coincide in one point of the excitation interval. In order to obtain regular selfpatch contributions we must calculate the contributions for an observation point just above the strip and then move that observation point to the matching point on the strip. This second step is a limit operation and was already discussed above in quite some detail.

In this section we will consider only the selfpatch contributions explicitly involving the singular behaviour of the spatial Green's dyadic. This singular behaviour was extensively discussed in Section 7.6. We will illustrate our approach by determining the contribution to the y-component of the tangential electric field at the edge of a strip excited by the dominant terms of the longitudinal (7.7) and transversal (7.8) current expansion near the edge. This means that we will study the contributions (7.11) for two particular cases. In both cases subscript a is equal to y as we study the contribution of the y-component of the electric field. In the first case subscript b is equal to x and the basis function $t(y')$ becomes $t_{1,j}(y')$ (see (7.10)). In the second case subscript b is equal to y and the basis function $t(y')$ becomes $t_{2,j}(y')$ (see (7.10)). The observation interval and the excitation interval coincide, that is, $z_e = z_o$ and they are selected to be the outermost interval at the left of strip j, that is, $y_{l,j} \leqslant y' \leqslant (y_{l,j} - \Delta_j)$. The matching point is located at the edge of the strip, that is, $y = y_{l,j}$. Consequently, the two integrals under study are given by (see also (7.7) and (7.8)):

$$I_1 = \lim_{z \to z_o} A_j \int_{y_{l,j}}^{y_{l,j}+\Delta_j} G_{yx}(y_{l,j}, z, y', z_e) \frac{1}{\sqrt{y' - y_{l,j}}} dy'$$

$$I_2 = \lim_{z \to z_o} B'_j \int_{y_{l,j}}^{y_{l,j}+\Delta_j} G_{yy}(y_{l,j}, z, y', z_e) \sqrt{y' - y_{l,j}} \, dy'.$$
(7.30)

We have selected these particular selfpatch contributions because they are the only ones which result in a singular field contribution even when we perform the limit operation after the integration in (7.30).

The further discussion will be focused on the contribution of the dominant singular terms of each of the components of G_{yx} and G_{yy} in (7.30). To this end we use the expressions for G_{yx} and G_{yy} given in (7.22). In (7.22) the singular contribution to G_{xy} comes from I_{TM}^1 and the singular contribution to G_{yy} comes from I_{TM}^2. Moreover, to obtain the singular part it suffices to

158   7 THIN STRIP TRANSMISSION LINE STRUCTURES

replace G_{TM} in the definition of I_{TM}^1 and I_{TM}^2 as given by (7.21) by G_{TM}^d defined by (7.24) and to take the dominant terms for large $|k_y|$ values. Finally, in the inverse Fourier transformation in (7.21) only the parts over $]-\infty, -k_c]$ and $[k_c, +\infty[$ have to be retained. For the singular part associated with I_1 in (7.30) this leads to

$$I_1 = \lim_{z \to z_0} a_j \int_{y_{l,j}}^{y_{l,j}+\Delta_j} \frac{1}{\sqrt{y' - y_{l,j}}} dy' \int_{k_c}^{+\infty} \sin[k_y(y' - y_{l,j})] e^{-k_y(z-z_e)} dk_y$$

(7.31)

while the singular contribution to I_2 becomes

$$I_2 = \lim_{z \to z_0} b'_j \int_{y_{l,j}}^{y_{l,j}+\Delta_j} \sqrt{y' - y_{l,j}} \, dy' \int_{k_c}^{+\infty} k_y \cos[k_y(y' - y_{l,j})] e^{-k_y(z-z_e)} dk_y.$$

(7.32)

To obtain (7.31) we used (7.27) for $m = 0$ ($p = 1$, TM case). To obtain (7.32) we used (7.27) for $m = 1$ ($p = 2$, TM case). The constants a_j and b'_j differ from A_j and B'_j in (7.30). The actual values of these coefficients is not calculated here. We refer the reader to [13] for a detailed calculation of these values. The integrals (7.31) and (7.32) can be calculated analytically by interchanging the two integrations. The results are

$$I_1 = a_j \left[\lim_{z \to z_0} \frac{\pi}{\sqrt{2(z - z_0)}} - \frac{2\cos(k_c \Delta_j)}{\sqrt{\Delta_j}} - 2\sqrt{2\pi k_c} \, S(k_c \Delta_j) \right] \quad (7.33)$$

and

$$I_2 = b'_j \left[\lim_{z \to z_0} -\frac{\pi}{2\sqrt{2(z - z_e)}} + \frac{2\cos(k_c \Delta_j)}{\sqrt{\Delta_j}} + \sqrt{2\pi k_c} \, S(k_c \Delta_j) \right]. \quad (7.34)$$

$S(k_c \Delta_j)$ represents the Fresnel sine integral [20]. The square root singularity in (7.33) and (7.34) does not imply that we have singular contributions to the integral equation. The singular terms correspond to the singularity of a field component in the vicinity of the edge of a metal plate [17]. Although the tangential electric field is zero everywhere on the strip this is not the case just above the strip. Both the longitudinal and transversal current component give rise to an electric field component which in fact becomes infinite just above the edges of the strip as is required by Meixner's edge condition [17]. For the actual selfpatch contribution the singular square root terms in (7.33) and (7.34) must be dropped. These non-regular terms correspond to the static field in the neighbourhood of the edge of the strip.

7.8 Case studies

In this section we illustrate our microstrip full-wave simulator by the analysis of the circuit parameters of the fundamental modes of single and coupled, coplanar and non-coplanar structures. We have also compared some of the results presented in this section with results obtained by other researchers which use the spectral domain approach. The comparisons made here show that our method presents a worthwhile alternative to the spectral domain techniques.

All case studies are restricted to non-magnetic material, that is, the relative permeability of each layer is equal to 1. In Chapter 3 we provided an example of two coupled non-coplanar microstrips, so this will not be repeated here.

7.8.1 *A single microstrip*

In this first example, which was also used in Chapter 2, we consider a single microstrip on top of a substrate with relative permittivity $\varepsilon_r = 11.7$. The substrate has a thickness $d = 3.17$ mm while the width of the strip is $w = 3.04$ mm. The microstrip and the relevant physical parameters are shown on Fig. 7.9. This typical configuration was also simulated by several other authors using a spectral domain technique [3], [4], and [10]. In our simulation the frequency was swept from d.c. to 15 GHz.

The effective relative permittivity $\varepsilon_{r,\text{eff}}$ and the characteristic impedance were calculated as a function of frequency. These two physical quantities suffice to construct a transmission line circuit equivalent of the microstrip. Figure 7.10 shows the simulated results. Ten basis functions were used in the determination of the circuit parameters.

The effective relative permittivity is an increasing function of frequency. This can be explained physically: as the frequency is increased the electromagnetic field becomes more and more confined in the substrate which results in a higher 'average' or effective permittivity. Our results are in excellent agreement with the results obtained by Kobayashi [10] who simulated the structure up to 10 GHz.

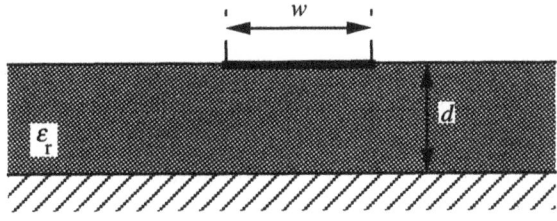

Fig. 7.9. A single microstrip.

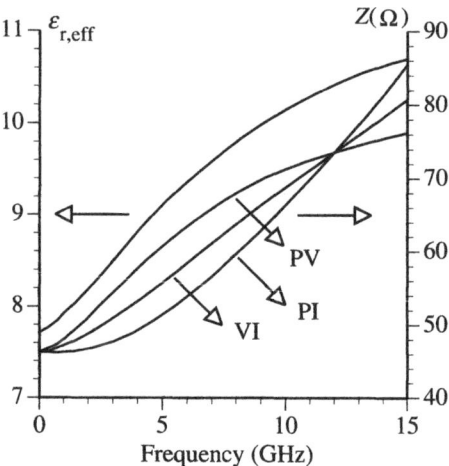

Fig. 7.10. Effective relative permittivity and characteristic impedance (according to the PI, PV, and VI definition) as a function of frequency.

Figure 7.10 also shows the characteristic impedances calculated according to the three definitions introduced in Chapter 2: a power–current, a power–voltage, and a voltage–current definition. The current is defined as the total longitudinal current flowing on the strip in the cross-section $x = 0$. The voltage is the voltage difference between the ground plane and the strip calculated along the symmetry axis of the structure. The frequency dependence is different for the three impedance definitions. However, the results coincide in the low-frequency limit as is predicted by the theory because in this limit the field is quasi-static and the longitudinal field becomes negligible. This low-frequency behaviour of the characteristic impedance is an indication of the accuracy of our simulation technique. In [5] a comparison is made between the power–current and the voltage–current definition. In the low-frequency limit the two definitions predict a different value for the characteristic impedance. This difference is a consequence of the inaccurate modelling of the current and is physically incorrect.

The characteristic impedance is also an increasing function of frequency. An explanation can be given for this behaviour. The dielectric confinement results in an increase of the effective permittivity and in a reduction of the effective width. The first effect increases the microstrip capacitance while the latter decreases the capacitance. Of the two competing effects the reduction of the effective width has the upper hand as the impedance is an increasing function of the frequency.

Figure 7.11 shows the variation of the longitudinal and transversal current as a function of the position on the strip and of the frequency. Because the longitudinal current is symmetric and the transversal current is antisym-

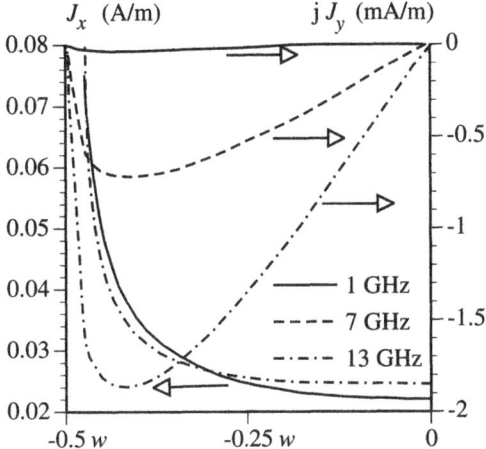

Fig. 7.11. Longitudinal current J_x at 1 and 13 GHz and transversal current J_y at 1, 7, and 13 GHz.

metric, Fig. 7.11 only displays the current components on the left half of the strip. For the displayed current profiles we used twenty intervals across the width of the strip. The total longitudinal current is kept constant. As the frequency is increased the longitudinal current moves to the edges and the current becomes constant over a larger portion of the strip while the singular behaviour becomes confined to the close vicinity of the two edges. The amplitude of the transversal current is an increasing function of frequency. At d.c. there is no transversal current but only a longitudinal current. Even at 13 GHz the transversal current is still much smaller than the longitudinal current. Figure 7.11 clearly shows that our simulation tool gives a very accurate prediction of the current variation as a function of the frequency.

7.8.2 *A suspended substrate line*

In the second case study we discuss the three-layered suspended substrate line shown in Fig. 7.12. The physical parameters are shown on the figure. In our simulation d takes the value 3 mm and $\varepsilon_r = 9.7$. The structure was also analysed by Jansen [6]. The full lines in Figs 7.13 and 7.14 show the simulated results for the effective relative permittivity and the characteristic impedance, respectively, obtained with the technique presented in this chapter. The dots are the results obtained by Jansen with a spectral domain technique. Both the effective relative permittivity and the characteristic impedance are in perfect agreement up to 8 GHz, the maximum frequency in [6]. We used ten divisions in our simulation.

162   7 THIN STRIP TRANSMISSION LINE STRUCTURES

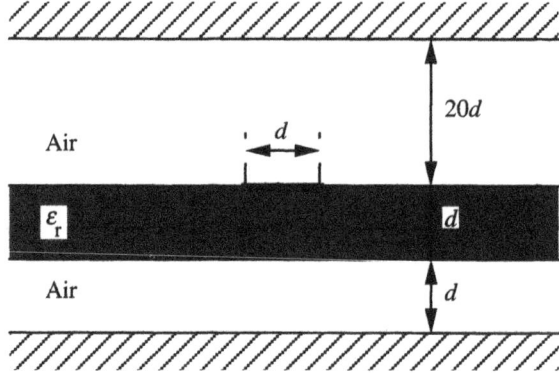

Fig. 7.12. A suspended substrate line.

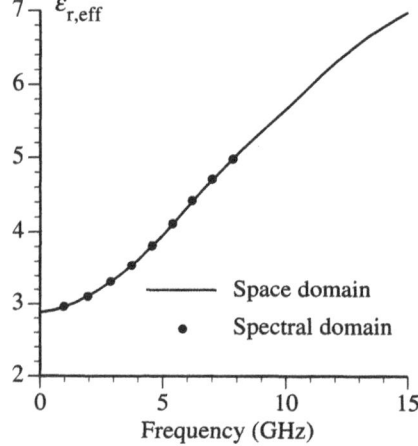

Fig. 7.13. Effective relative permittivity of the fundamental eigenmode of the structure of Fig. 7.12.

7.8.3 *Two coupled symmetric microstrips*

In this case study we present the analysis of two coupled symmetric microstrips. The structure is shown in Fig. 7.15. This structure propagates two fundamental modes, an even and an odd mode. Both strips have a width $w_1 = w_2 = 0.6$ mm. The substrate thickness is $d = 0.64$ mm and the relative permittivity of the substrate is $\varepsilon_r = 9.9$. The distance s between the two strips is a parameter and takes the values 0.02 mm, 0.6 mm, and 2.0 mm. In the limit $s = \infty$ both strips no longer couple and we find the circuit parameters of the single strip.

Figure 7.16 shows the effective relative permittivity of both modes. The

7.8 CASE STUDIES   163

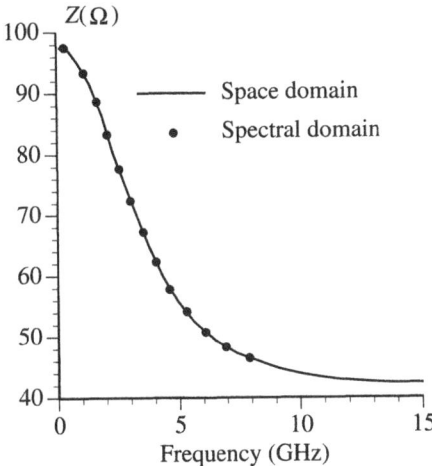

Fig. 7.14. Characteristic impedance of the fundamental eigenmode of the structure of Fig. 7.12.

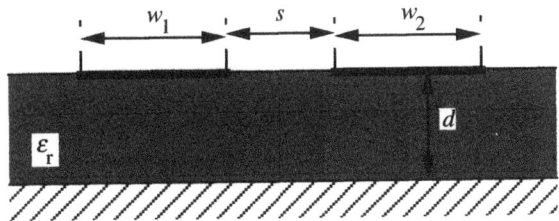

Fig. 7.15. Geometry of two coupled symmetric microstrips.

full lines are our simulated results. The dashed lines are the results obtained by Jansen using a spectral domain technique [6]. The results are in very good agreement. The effective permittivity of the even mode is higher than that of the odd mode for the same distance between the strips. This is a consequence of the fact that a larger portion of the even mode field propagates in the substrate compared to the odd mode.

Next, Fig. 7.17 displays the characteristic impedances of the even and odd modes. We note that for each mode both strips have the same characteristic impedance because of the symmetry in the structure. Our results (full lines) for the characteristic impedances have also been compared with the results obtained by Jansen (dashed lines). Unlike the results for the effective permittivity, for the characteristic impedances there is a clear discrepancy for higher frequencies between our results and Jansen's. This is a consequence of the different definition applied for the characteristic impedance. Our results are obtained with the power–current definition while Jansen used a power–voltage definition. Both definitions produce results that are in good

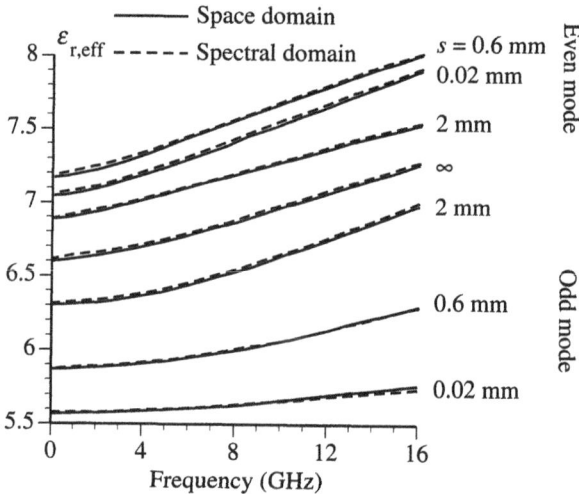

Fig. 7.16. Effective permittivity of the even and odd modes of the symmetric strip structure of Fig. 7.15.

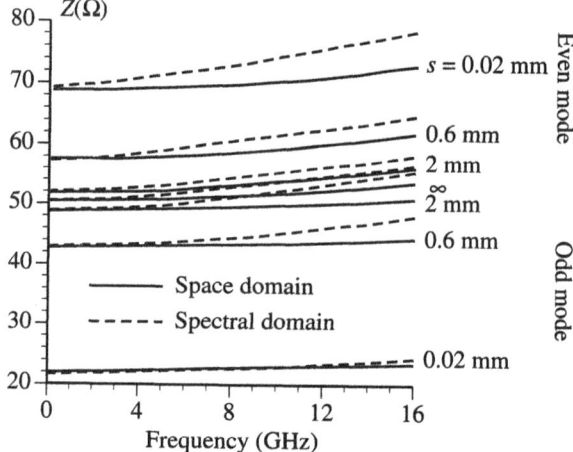

Fig. 7.17. Characteristic impedances for the even and odd modes of the symmetric strip structure of Fig. 7.15.

agreement for low frequencies. However, Fig. 7.17 shows that the impedances based upon the power–current definition keep their static value over a wider frequency range.

7.8.4 Two coupled asymmetric microstrips

As a second case study of coupled microstrips we have analysed the asymmetric microstrip configurations of Fig. 7.18. The widths of the two

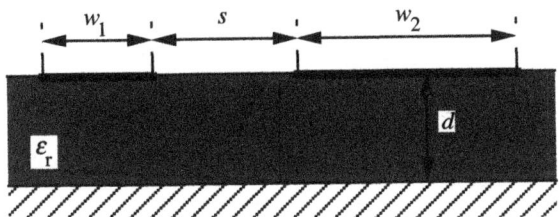

Fig. 7.18. Geometry of two coupled asymmetric microstrips.

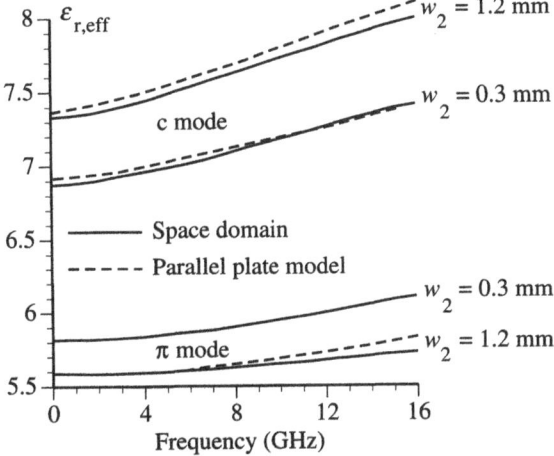

Fig. 7.19. Effective permittivities of the c and π modes for both strips of the asymmetric strip structure of Fig. 7.18.

strips are $w_1 = 0.6$ mm and $w_2 = 0.3$ mm or $w_2 = 1.2$ mm, respectively, and the distance between the strips is $s = 0.4$ mm. The thickness of the substrate is 0.635 mm and the relative permittivity $\varepsilon_r = 9.9$. The present configuration was also analysed by Tripathi [9] using a parallel plate model.

The structure has two fundamental modes: a c and π mode. For a symmetric structure the c mode corresponds to the even mode and the π mode to the odd mode.

The effective permittivities of the c and π modes for both strips obtained with our spatial technique and the parallel plate method of Tripathi are compared on Fig. 7.19. In [9] no results are presented for the characteristic impedances. Therefore Fig. 7.20 only shows the characteristic impedances of both strips for both modes obtained with our technique. The variation of the impedances as a function of frequency is very small. The impedances keep their static value over a wider frequency range than the effective permittivities.

166   7 THIN STRIP TRANSMISSION LINE STRUCTURES

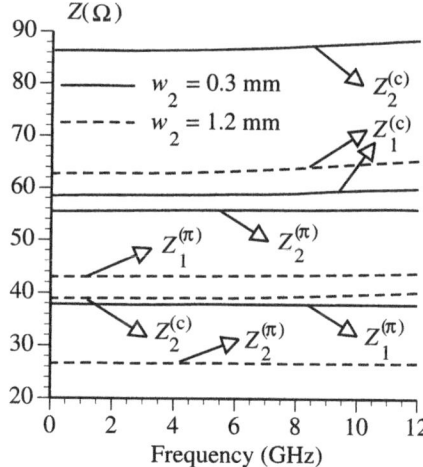

Fig. 7.20. Characteristic impedances of the c and π modes for both strips of the asymmetric strip structure of Fig. 7.18.

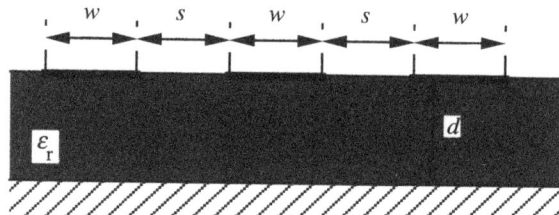

Fig. 7.21. Geometry of a three coupled line structure.

7.8.5 Three coupled lines

We conclude this section on the case studies with the presentation of the current profiles of the three fundamental modes of a three coupled line structure. The structure is given in Fig. 7.21. Each strip has a width $w = 1$ mm and the distance between two adjacent strips is $s = 1$ mm. The substrate thickness is $d = 1$ mm and the relative permittivity of the substrate is $\varepsilon_r = 9.9$. Figure 7.22 displays the longitudinal currents of the three fundamental modes: two even and one odd mode at 1 GHz.

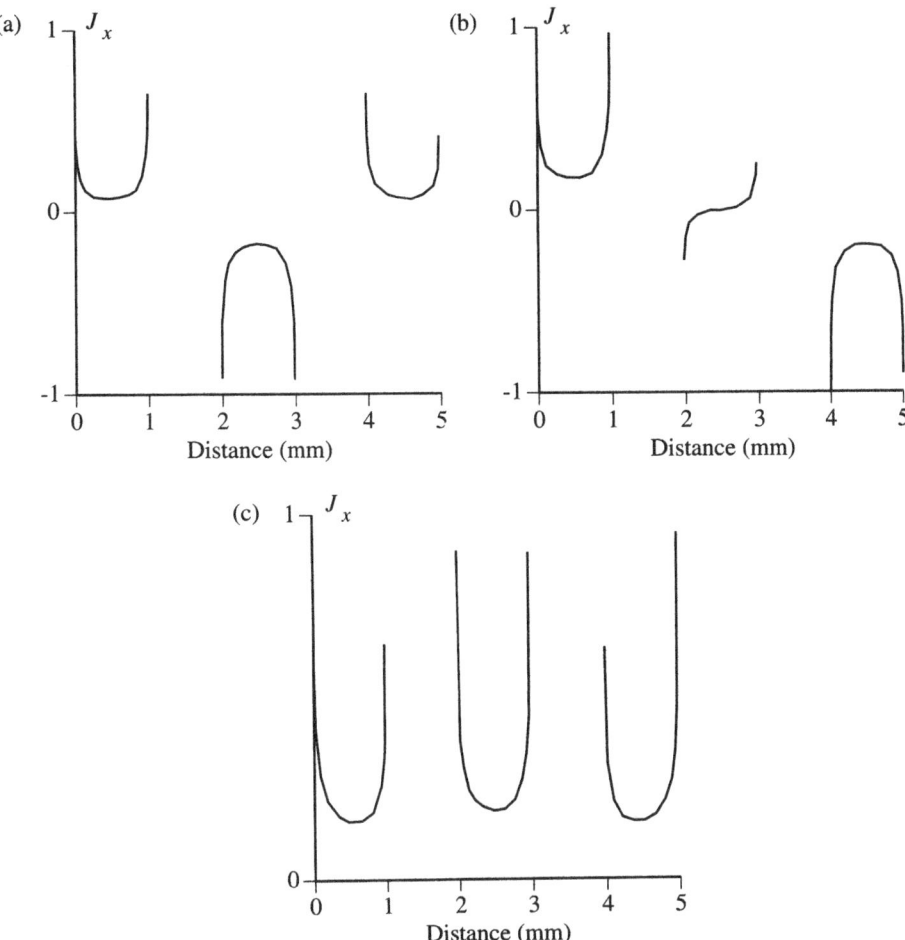

Fig. 7.22. Longitudinal currents of the three fundamental modes of the structure of Fig. 7.21 at 1 GHz: (a) first even mode; (b) first odd mode; (c) second even mode.

References

[1] Delbare, W. and De Zutter, D. (1989). Space domain Green's function approach to the capacitance calculation of multiconductor lines in multilayered dielectrics with improved surface charge modelling. *IEEE Transactions on Microwave Theory and Techniques*, **MTT-37**, 10, 1562–1568.

[2] Olyslager, F., Faché, N., and De Zutter, D. (1991). New fast and accurate line parameter calculation of general multiconductor transmission lines in multi-layered media. *IEEE Transactions on Microwave Theory and Techniques*, **MTT-39**, 6, 901–909.

[3] Denlinger, E. J. (1971). A frequency dependent solution for microstrip transmission lines. *IEEE Transactions on Microwave Theory and Techniques*, **MTT-19**, 1, 30–39.

[4] Itoh, T. and Mittra, R. (1973). Spectral-domain approach for calculating the dispersion characteristics of microstrip lines. *IEEE Transactions on Microwave Theory and Techniques*, **MTT-21**, 7, 496–499.

[5] Knorr, J. B. and Tufekcioglu, A. (1975). Spectral-domain calculations of microstrip characteristic impedance. *IEEE Transactions on Microwave Theory and Techniques*, **MTT-23**, 9, 725–728.

[6] Jansen, R. H. (1979). Unified user-oriented computation of shielded, covered and open planar microwave and millimeter-wave transmission-line characteristics. *Microwaves, Optics and Acoustics*, **3**, 1, 14–22.

[7] Fukuoka, Y., Zhang, Q., Neikirk, D., and Itoh, T. (1985). Analysis of multilayer interconnection lines for a high-speed digital integrated circuit. *IEEE Transactions on Microwave Theory and Techniques*, **MTT-23**, 6, 527–532.

[8] Jansen, R. H. (1985). The spectral domain approach for microwave integrated circuits. *IEEE Transactions on Microwave Theory and Techniques*, **MTT-33**, 10, 1043–1056.

[9] Tripathi, V. K. (1986). A dispersion model for coupled microstrips. *IEEE Transactions on Microwave Theory and Techniques*, **MTT-34**, 1, 66–71.

[10] Kobayashi, M. and Ando, F. (1987). Dispersion characteristics of open microstrip lines. *IEEE Transactions on Microwave Theory and Techniques*, **MTT-35**, 2, 101–105.

[11] Das, N. and Pozar, D. (1987). A generalized spectral-domain Green's function for multilayered dielectric substrates with application to multilayered transmission lines. *IEEE Transactions on Microwave Theory and Techniques*, **MTT-35**, 3, 326–335.

[12] Marques, R. and Horno, M. (1987). On the spectral Green's function for stratified linear media. *Proceedings of the Institute of Electrical Engineers*, **134**, H, 3, 241–248.

[13] Faché, N. and De Zutter, D. (1988). Rigorous full-wave space-domain solution for dispersive microstrip lines. *IEEE Transactions on Microwave Theory and Techniques*, **MTT-36**, 4, 733–737.

[14] Faché, N. and De Zutter, D. (1989). Circuit parameters for single and coupled microstrip lines by a rigorous full-wave space domain analysis. *IEEE Transactions on Microwave Theory and Techniques*, **MTT-37**, 2, 421–425.

[15] Faché, N., Van Hese, J., and De Zutter, D. (1989). Generalised space domain Green's dyadic for multilayered media with special application to microwave interconnections. *Journal of Electromagnetic Waves and Applications*, **3**, 7, 651–669.

[16] Harrington, R. F. (1968). *Field Computations by Moment Methods*. Macmillan, New York.

[17] Meixner, J. (1972). The behaviour of electromagnetic fields at edges. *IEEE Transactions on Antennas and Propagation*, **AP-20**, 7, 442–446.

[18] Felsen, L. B. and Marcuvitz, N. (1973). *Radiation and Scattering of Waves*. Prentice-Hall, Englewood Cliffs, NJ.

[19] Itoh, T. (ed.) (1989). *Numerical Techniques for Microwave and Millimeter-Wave Passive Structures*. Wiley, New York.

[20] Abramowitz, M. and Stegun, I. A. (1970). *Handbook of Mathematical Functions with Formulas, Graphs and Mathematical Tables*. Dover, New York.

8
LOSSLESS WIRE TRANSMISSION LINE STRUCTURES

8.1 Introduction

Wire transmission line structures are the second type of interconnection structures handled in this book. This type of transmission line, like the strip transmission line structures, allows a special treatment. Wire transmission line structures are gaining more and more practical importance in, for example, the Multiwire®, Microwire®, and Microdot® technologies.

Several authors have analysed wire structures in the quasi-TEM limit. In [1] a single wire in a medium consisting of two layers is studied. The authors use the Green's function of the layered medium and replace the wire with a number of line charges located inside the wire. In [2] a wire is approximated by a regular polygonal waveguide and the layered medium is handled with polarization charges. A more recent and accurate method is described in [3]. This paper also handles polygonal conductors in multilayered media but the Green's function of the layered medium is used and a more accurate representation of the charge density on the conductors, as compared to [2], is introduced. This method gives accurate results only if a sufficient number of sides of the polygons is considered. Care must be taken when wires are closely spaced together or are close to a ground plane. A more accurate and elegant method is presented by the present authors in [4]. In this work curved boundaries are explicitly taken into account. Hence, it is not necessary to approximate wires by polygonal conductors.

In this chapter a full-wave analysis of wire transmission line structures is presented. Papers by the present authors, who were the first to handle the full-wave analysis of wires, can be found in [5], [6], [7], and [8]. Other authors, in [9] and [10], have considered wire transmission lines as a special case of the full-wave analysis of open microstrip lines of arbitrary cross-section. Here again, the wires are approximated by a polygonal boundary. In the next chapter, it will be explained that this can indeed be done with good success.

As in Chapter 7 for infinitely thin strips, our purpose is the determination of the circuit parameters of the modes propagated by a general wire transmission line structure. The sources of a mode are the surface currents on the wires. They are the unknowns of the problem. These surface currents are solutions to a one-dimensional boundary integral equation at the

8 LOSSLESS WIRE TRANSMISSION LINE STRUCTURES

boundaries of the wires. The method of Galerkin is applied to solve this integral equation which immediately results in the propagation constants and surface currents of the modes. From the surface currents the power propagated by the modes is calculated.

We start with a description of the geometry of the wire transmission line structures. Secondly, the integral equation is constructed in which the fields are decomposed into incoming and scattered parts. Then an overview of the calculation of these incoming and scattered fields is presented, followed by a detailed calculation. Finally the longitudinal currents and the propagated powers are calculated. The whole theory is illustrated with some case studies in the last section.

8.2 Geometry

The geometry of the cross-section of the wire transmission line structure is shown in Fig. 8.1. The general multilayered medium counts L layers in which D perfectly conducting circular wires are embedded. The subscript j ($j = 1, \ldots, D$) is used to refer to the jth wire. Each wire must be embedded in a single layer and is not allowed to cross any layer interfaces. As in Chapter 6 several types of layered media (open, semi-open, and closed) are considered. Figure 8.1 gives an example of a semi-open medium, that is, microstrip type. When a wire is located in a semi-infinite layer an additional fictitious layer interface is introduced to make sure that every wire is embedded in a finite layer. There are D_i wires embedded in layer i ($i = 1, \ldots, L$) and the subscript

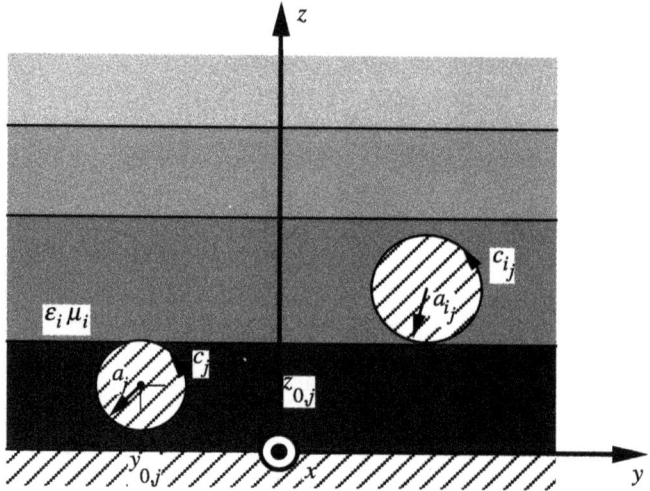

Fig. 8.1. Geometry of a wire transmission line structure.

8.3 CONSTRUCTION OF THE INTEGRAL EQUATION

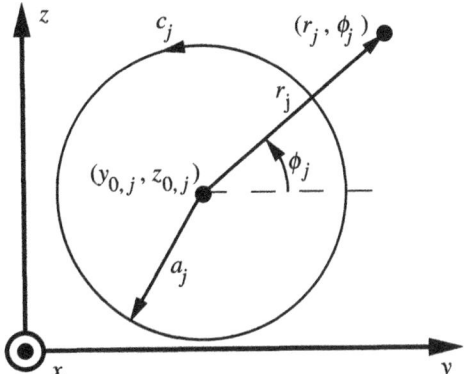

Fig. 8.2. Local polar coordinate system (r_j, ϕ_j, x) associated with wire j.

i_j ($i_j = 1, \ldots, D_i$) refers to the i_jth wire in layer i. The total number of wires, D, is equal to the sum of the number of wires D_i in each layer i.

Wire j has a radius a_j and its centre point is located at $(y_{0,j}, z_{0,j})$. With each wire j we associate a local polar coordinate system (r_j, ϕ_j, x) originating at the centre of the wire. This coordinate system is shown in Fig. 8.2. The boundary of wire j is denoted by c_j. The radius and the boundary of the jth wire in layer i are denoted by a_{i_j} and c_{i_j} respectively. These more complicated double subscripts will only be needed in the general outline of the theory in the next two sections. In the remaining sections a simplified notation without subscripts or with a single subscript, is used.

8.3 Construction of the integral equation

The modes of the wire transmission line structure are described by an integral equation. The solution of this integral equation allows one to determine the parameters in the equivalent circuit model. Indeed, the solution of the integral equation will result in the propagation constants β of the modes and the modal surface currents on each wire. From these modal surface currents one calculates the total longitudinal current on each wire and the modal field distributions. The propagated powers of all the modes are found by combining and integrating these modal field distributions.

8.3.1 *General representation of the modal fields*

Because we are working with modal fields and currents we will drop the common time and longitudinal dependence $e^{j(\omega t - \beta x)}$ in all field and current

172   8 LOSSLESS WIRE TRANSMISSION LINE STRUCTURES

components. The surface current $\mathbf{J}_j$ on each wire has two components $J_{x,j}(\phi_j)$ and $J_{\phi,j}(\phi_j)$.

The total modal fields $[\mathbf{E}(\mathbf{r}), \mathbf{H}(\mathbf{r})]$ ($\mathbf{r} = y\mathbf{1}_y + z\mathbf{1}_z$) are generated by the surface currents on each wire. The fields generated by the surface currents on one wire will consist of the combination of incoming fields and scattered fields.

In our way of working, the incoming fields of wire i_j are defined only in the layer i surrounding this wire. They are equal to the fields that would be generated by the surface current on wire i_j if this wire were to be placed in a homogeneous space with the same material properties as layer i. These incoming fields are denoted by $[\mathbf{E}^{in}_{i_j}(\mathbf{r}), \mathbf{H}^{in}_{i_j}(\mathbf{r})]$. The superscript 'in' refers to the fact that we are dealing with incoming fields and the subscript i_j explains in which layer these fields exist and which wire excites these fields.

The scattered fields arise from the fact that wire i_j is not in homogeneous space but in a layered medium. The incoming fields will penetrate into and will be scattered by all the layers of the structure. In the following the term *scattered fields* is used for the fields that penetrate into those layers that are different from the excitation layer i and for the reflected fields in the excitation layer i. The scattered fields in all layers due to wire i_j are represented by $[\mathbf{E}^{sc}_{i_j}(\mathbf{r}), \mathbf{H}^{sc}_{i_j}(\mathbf{r})]$. The superscript 'sc' now refers to the fact that we are dealing with scattered fields. If we are not explicitly referring to the layer in which wire j is located, we will denote the scattered fields due to that wire by $[\mathbf{E}^{sc}_j(\mathbf{r}), \mathbf{H}^{sc}_j(\mathbf{r})]$.

Finally, the total modal fields generated by the surface currents on all the wires are given by

$$\mathbf{E}(\mathbf{r}) = \sum_{i=1}^{L} \sum_{i_j=1}^{D_i} \mathbf{E}^{in}_{i_j}(\mathbf{r}) + \sum_{j=1}^{D} \mathbf{E}^{sc}_j(\mathbf{r})$$

$$\mathbf{H}(\mathbf{r}) = \sum_{i=1}^{L} \sum_{i_j=1}^{D_i} \mathbf{H}^{in}_{i_j}(\mathbf{r}) + \sum_{j=1}^{D} \mathbf{H}^{sc}_j(\mathbf{r}).$$

(8.1)

We have summed all incoming and scattered fields.

8.3.2 *The integral equation*

The modal fields $[\mathbf{E}(\mathbf{r}), \mathbf{H}(\mathbf{r})]$ (8.1) are a solution to Maxwell's equations and the corresponding boundary conditions. Due to the modal longitudinal $e^{-j\beta x}$ dependence these equations reduce to the two-dimensional field problem described in Appendix A. From this appendix we know that all the field components can be calculated from the longitudinal field components $[E_x(\mathbf{r}), H_x(\mathbf{r})]$. Consequently these field components will act as basis components in the representation of the modal fields.

In the representation (8.1) of the modal fields, the surface currents $\mathbf{J}_j$ are

8.4 SOLUTION OF THE INTEGRAL EQUATION

the actual unknowns of the problem. In order to construct an integral equation for these currents, we have to incorporate the boundary conditions at the perfect conductors. It is sufficient to impose the boundary conditions for the longitudinal fields. The other boundary conditions will be fulfilled automatically.

The longitudinal electric field E_x vanishes at the wire boundaries because it is tangential to the perfectly conducting surfaces. The tangential magnetic field however follows directly from the surface current on the wire. Indeed, at the wire surface of wire j the following relation between the surface current and the magnetic field exists:

$$\mathbf{1}_{r_j} \times \mathbf{H}(\mathbf{r}_j) = \mathbf{J}_j. \tag{8.2}$$

The unit vector $\mathbf{1}_{r_j}$ in (8.2) is the radially directed outward normal to the wire surface. From (8.2), the relevant boundary condition for the longitudinal magnetic field is

$$H_x(\mathbf{r}_j) = -J_{\phi_j, j}. \tag{8.3}$$

From these boundary conditions for the longitudinal fields we construct the integral equation in the following way. We determine or 'observe' the value of the total longitudinal fields on each wire. Let $\mathbf{r}_{p_o} = a_{p_o} \cos(\phi_{p_o})\mathbf{1}_y + a_{p_o} \sin(\phi_{p_o})\mathbf{1}_z$ represent a point on the surface of such an observation wire p_o ($p_o = 1, \ldots, D_p$) in layer p ($p = 1, \ldots, L$). Then these total longitudinal fields satisfy:

$$\lim_{\mathbf{r} \gtrless \mathbf{r}_{p_o}} \left[\sum_{j=1}^{D_p} E_{x, p_j}^{\text{in}}(\mathbf{r}) + \sum_{j=1}^{D} E_{x, j}^{\text{sc}}(\mathbf{r}) \right] = 0$$

$$\lim_{\mathbf{r} \gtrless \mathbf{r}_{p_o}} \left[\sum_{j=1}^{D_p} H_{x, p_j}^{\text{in}}(\mathbf{r}) + \sum_{j=1}^{D} H_{x, j}^{\text{sc}}(\mathbf{r}) \right] = J_{\phi_{p_o}, p_o}. \tag{8.4}$$

The longitudinal boundary conditions (8.4) are expressed on the circumference of the observation wire p_o. The notation $\mathbf{r} \gtrless \mathbf{r}_{p_o}$ indicates that the surface of wire p_o is approached from the outside. In the following we will show that all the fields in (8.4) can in fact be expressed in terms of integrals over the unknown surface currents J_{x, p_o} and $J_{\phi_{p_o}, p_o}$ on the observation wire p_o. This implies that equations (8.4) are in fact two coupled integral equations in the two unknown components of the surface currents, that is, the ϕ- and x-component. We will refer to (8.4) as 'the integral equation'.

8.4 Solution of the integral equation

The integral equation (8.4) is solved by means of the method of Galerkin. In the previous chapter a point matching approach was used. The method

8 LOSSLESS WIRE TRANSMISSION LINE STRUCTURES

of Galerkin means that we use test functions to test the integral equation which are equal to the basis functions used in the representation of the unknown surface currents. The complex amplitudes of the basis functions in the representation of $\mathbf{J}_j$ form a set of discrete unknowns. By imposing the equality of the scalar product of the right- and left-hand sides of (8.4) with each test function we obtain a linear system of equations with the basis function amplitudes as unknowns. This linear system forms an eigenvalue problem with the propagation constants β of the modes of the original problem as eigenvalues.

Due to the cylindrical shape of the wires, a representation of the surface current in terms of a Fourier series in ϕ_j will emerge in a very natural way. For the actual calculations these infinite Fourier series will be limited to a finite number of terms. This in fact means that the basis and test functions used in the present chapter are $e^{jn\phi_j}$ ($n = 0, \pm 1, \pm 2, \ldots$).

In the rest of this section the calculation of the incoming and scattered fields is outlined. Detailed calculations are presented in subsequent sections. This allows the reader to get an overall view of the way the problem is handled without being confused by too many additional mathematical details. The proposed calculation technique for the incoming fields requires the replacement of the surface currents as unknowns in (8.4) by some new quantities on the wire surfaces. These new quantities are equivalent to the surface currents.

8.4.1 The incoming fields

The incoming fields associated with wire i_j in layer i are equal to the fields generated by the surface currents $\mathbf{J}_{i_j}$ when this wire is in homogeneous space with the same permeability and permittivity as layer i.

In Appendix A the following integral representation for the longitudinal incoming fields is obtained:

$$E^{\text{in}}_{x,i_j}(\mathbf{r}) = \int_0^{2\pi} G_i \frac{\partial E_x}{\partial r'_{i_j}} a_{i_j} \, d\phi_{i_j}$$

$$H^{\text{in}}_{x,i_j}(\mathbf{r}) = -\int_0^{2\pi} H_x \frac{\partial G_i}{\partial r'_{i_j}} a_{i_j} \, d\phi_{i_j} \tag{8.5}$$

with $G_i(\mathbf{r}|\mathbf{r}')$ the Green's function of homogeneous space with the same properties as layer i. This function is given by

$$G_i(\mathbf{r}|\mathbf{r}') = \frac{j}{4} H_0^{(2)}(\gamma_i |\mathbf{r} - \mathbf{r}'|) \tag{8.6}$$

8.4 SOLUTION OF THE INTEGRAL EQUATION

with

$$\gamma_i^2 = \omega^2 \varepsilon_i \mu_i - \beta^2 \tag{8.7}$$

and with $H_0^{(2)}$ the Hankel function of the second kind and zeroth order. The integration in (8.5) extends over the circumference of wire i_j. The longitudinal incoming electric field follows from the normal derivative of the total longitudinal electric field E_x at the wire surface. On the other hand the longitudinal incoming magnetic field is determined by the total longitudinal magnetic field H_x at the surface. Consequently we will use the radial derivative of the longitudinal electric field and the total longitudinal magnetic field at the wire boundaries as new unknowns. The original unknowns, that is, the currents J_{x,i_j} and J_{ϕ_{ij},i_j} at wire i_j, are related to the new unknowns in the following way (cf. Appendix A):

$$H_x(\mathbf{r}_{i_j}) = -J_{\phi_{ij},i_j}$$

$$\frac{\partial E_x(\mathbf{r}_{i_j})}{\partial r_{i_j}} = -\frac{\gamma_i^2}{j\omega \varepsilon_i} H_{\phi_{ij}} - \frac{\beta}{\omega \varepsilon_i a_{i_j}} \frac{\partial H_x}{\partial \phi_{ij}} = -\frac{\gamma_i^2}{j\omega \varepsilon_i} J_{x,i_j} + \frac{\beta}{\omega \varepsilon_i a_{i_j}} \frac{\partial J_{\phi_{ij},i_j}}{\partial \phi_{ij}}. \tag{8.8}$$

Due to this change in unknowns the right-hand side of the second equation in (8.4) has to be replaced by the total longitudinal magnetic field.

The unknown functions at the wire surfaces are expanded in an angular Fourier series. The amplitudes in this expansion are the discrete unknowns of the eigenvalue problem. Thus the basic unknowns on wire surface j are expanded as follows:

$$\frac{\partial E_x(\mathbf{r}_j)}{\partial r_j} = \sum_{n=-\infty}^{+\infty} A_{n,j} e^{jn\phi_j}$$

$$H_x(\mathbf{r}_j) = \sum_{n=-\infty}^{+\infty} B_{n,j} e^{jn\phi_j}. \tag{8.9}$$

The angular Fourier coefficients $A_{n,j}$ and $B_{n,j}$ of all wires form the eigenvector in the discrete eigenvalue problem. Equation (8.9) is an exact representation of the unknown functions. However, in a practical implementation we have to truncate the series in (8.9) to $2N + 1$ terms ($n = 0, \pm 1, \pm 2, \ldots \pm N$). In this case the representation is only an approximation. In many practical situations, such as a single wire not too close to a ground plane or a set of not too strongly coupled wires, the series (8.9) converges very fast and N will typically be smaller than 3. The expansion of the unknowns in a Fourier series is the first step in Galerkin's method. The set of basis functions used on each wire j is $e^{jn\phi_j}$ with $n = 0, \pm 1, \pm 2, \ldots \pm N$.

The longitudinal incoming fields are now calculated by substituting the expansions (8.9) into the expressions (8.5). After interchanging the integration and summation, the integration can be performed analytically. This is done in Section 8.5 and results in the following expansion in terms of Hankel functions:

$$E^{in}_{x,i_j}(\mathbf{r}) = \sum_{n=-\infty}^{+\infty} C_{n,i_j} H_n^{(2)}(\gamma_i r_{i_j}) e^{jn\phi_{i_j}}$$

$$H^{in}_{x,i_j}(\mathbf{r}) = \sum_{n=-\infty}^{+\infty} D_{n,i_j} H_n^{(2)}(\gamma_i r_{i_j}) e^{jn\phi_{i_j}}.$$

(8.10)

The coefficients C_{n,i_j} (respectively D_{n,i_j}) are proportional to A_{n,i_j} (respectively B_{n,i_j}).

The incoming fields at the boundary of wire i_j are immediately obtained from (8.10) by setting r_{i_j} equal to a_{i_j}, that is, by replacing the general position vector $\mathbf{r}$ by the position vector $\mathbf{r}_{i_j}$ of a point on the boundary:

$$E^{in}_{x,i_j}(\mathbf{r}_{i_j}) = \sum_{n=-\infty}^{+\infty} C_{n,i_j} H_n^{(2)}(\gamma_i a_{i_j}) e^{jn\phi_{i_j}}$$

$$H^{in}_{x,i_j}(\mathbf{r}_{i_j}) = \sum_{n=-\infty}^{+\infty} D_{n,i_j} H_n^{(2)}(\gamma_i a_{i_j}) e^{jn\phi_{i_j}}.$$

(8.11)

In (8.11) the incoming fields on wire i_j are already expanded in the selected basis functions, so these fields are already cast into the proper form for the testing of the integral equation. This constitutes the second step in Galerkin's method. In our case this testing simply amounts to equating the left- and right-hand sides of the integral equation (8.4) on a per-Fourier-term basis. Due to the orthogonality of the terms of a Fourier series expansion, the above procedure yields the same result as multiplying the left- and right-hand sides with one of the basis functions $e^{jn\phi_{p_o}}$ and integrating over the circumference.

The incoming fields of wire i_j also contribute to the fields at the surface of the other wires in layer i. This contribution is calculated by transforming the incoming fields (8.10) to the spatial Fourier domain and propagating them to the observation wire i_o. Inverse Fourier transformation and angular Fourier decomposition brings us to the following expansion:

$$E^{in}_{x,i_j}(\mathbf{r}_{i_o}) = \sum_{n=-\infty}^{+\infty} E^{i_o}_{n,i_j} e^{jn\phi_{i_o}}$$

$$H^{in}_{x,i_j}(\mathbf{r}_{i_o}) = \sum_{n=-\infty}^{+\infty} F^{i_o}_{n,i_j} e^{jn\phi_{i_o}}.$$

(8.12)

The elaboration of (8.12) is presented in Section 8.6. The coefficients $E^{i_o}_{n,i_j}$ (respectively $F^{i_o}_{n,i_j}$) are proportional to A_{m,i_j} (respectively B_{m,i_j}).

8.4.2 The scattered fields

The scattered fields owe their origin to the fact that wire i_j is not placed in homogeneous space but is embedded in the layered structure. The incoming fields reflect in the layered medium and generate the scattered fields.

The scattered fields are calculated in the spectral domain. The spectral longitudinal incoming fields act as sources at the top and bottom of layer i in the TE and TM cascades of the layered medium (Fig. 8.3). The cascades are solved with the technique described in Chapter 6 and this results in the TE and TM components of the scattered fields. Combining these components and inverse Fourier transformation gives us the spatial longitudinal scattered fields. This inverse transformation requires a numerical integration.

For the last step in the method of Galerkin, an angular Fourier decomposition at the wire surfaces is required. The scattered fields generated by excitation wire j can be decomposed in the following Fourier series at the surface of the observation wire 'o':

$$E^{sc}_{x,j}(\mathbf{r}_o) = \sum_{n=-\infty}^{+\infty} G^o_{n,j} e^{jn\phi_o}$$

$$H^{sc}_{x,j}(\mathbf{r}_o) = \sum_{n=-\infty}^{+\infty} H^o_{n,j} e^{jn\phi_o}. \tag{8.13}$$

The coefficients $G^o_{n,j}$ and $H^o_{n,j}$ are linear combinations of all basis unknowns $A_{m,j}$ and $B_{m,j}$ ($m = 0, \pm 1, \pm 2, \ldots$) originally introduced in (8.9). In those coefficients the superscript 'o' refers to the observation wire 'o' and the second subscript j refers to the excitation wire j.

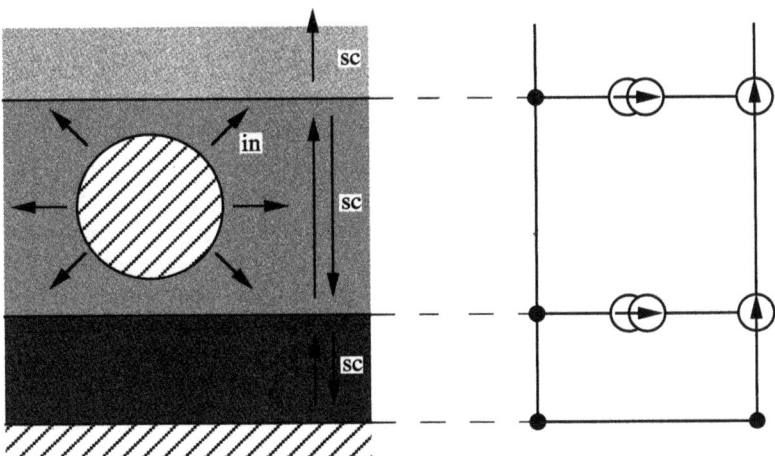

Fig. 8.3. Equivalent sources in the TE or TM cascades for the incoming fields.

8.4.3 Construction of the discrete eigenvalue problem

The total longitudinal fields on each wire i_o in layer i consist of the incoming fields of all wires in that particular layer, including wire i_o, and the scattered fields of all wires. From the expansions (8.11) and (8.12) for the incoming fields and (8.13) for the scattered fields we get the following expansions for the total fields at wire i_o:

$$E_x(\mathbf{r}_{i_o}) = \sum_{n=-\infty}^{+\infty} \left[C_{n,i_o} H_n^{(2)}(\gamma_i a_{i_o}) + \sum_{i_j=1; i_j \neq i_o}^{D_i} E_{n,i_j}^{i_o} + \sum_{j=1}^{D} G_{n,j}^{i_o} \right] e^{jn\phi_{j_o}}$$

$$H_x(\mathbf{r}_{i_o}) = \sum_{n=-\infty}^{+\infty} \left[D_{n,i_o} H_n^{(2)}(\gamma_i a_{i_o}) + \sum_{i_j=1; i_j \neq i_o}^{D_i} F_{n,i_j}^{i_o} + \sum_{j=1}^{D} H_{n,j}^{i_o} \right] e^{jn\phi_{j_o}}.$$

(8.14)

Finally we impose the boundary conditions at the wire surfaces for the total longitudinal fields. The total electric field vanishes at the wire surfaces. This means that the coefficients of the basis functions $e^{jn\phi_{j_o}}$ of the electric field in (8.14) also vanish due to the orthogonality of these functions. In both (8.9) and (8.14) expressions are given for the total magnetic field on the wire surface. Due to orthogonality, the expansions in (8.9) and (8.14) will be equal only if the coefficients of the same order are equal. The above conditions for the electric and magnetic field expansion coefficients lead to the following system of equations:

$$C_{n,i_o} H_n^{(2)}(\gamma_i a_{i_o}) + \sum_{i_j=1; i_j \neq i_o}^{D_i} E_{n,i_j}^{i_o} + \sum_{j=1}^{D} G_{n,j}^{i_o} = 0$$

$$D_{n,i_o} H_n^{(2)}(\gamma_i a_{i_o}) + \sum_{i_j=1; i_j \neq i_o}^{D_i} F_{n,i_j}^{i_o} + \sum_{j=1}^{D} H_{n,j}^{i_o} = B_{n,i_o}$$

(8.15)

$$n = 0, \pm 1, \pm 2, \ldots \quad i = 1, \ldots, L \quad i_o = 1, \ldots, D_i.$$

In (8.15) C_{n,i_o} is proportional to A_{n,i_o} while D_{n,i_o} is proportional to B_{n,i_o}. $E_{n,i_j}^{i_o}$ depends on all A_{m,i_j} values ($m = 0, \pm 1, \pm 2, \ldots$). In the same way $F_{n,i_j}^{i_o}$ depends on all B_{m,i_j} values. Finally $G_{n,j}^{i_o}$ and $H_{n,j}^{i_o}$ depend on all $A_{m,j}$s and $B_{m,j}$s. The discrete system (8.15) replaces the integral equation (8.4). In a practical implementation n is limited to $n = 0, \pm 1, \pm 2, \ldots \pm N$. As mentioned before N is typically smaller than 3. This implies that the integral equation is replaced by a homogeneous linear eigenvalue system of $2D(2N+1)$ equations with $2D(2N+1)$ unknown coefficients $A_{n,j}$ and $B_{n,j}$. Remember that D represents the total number of wires. The coefficients in this system are functions of the propagation constants β of the modes. These propagation constants are the eigenvalues of the system of equations and are found as zeros of the determinant of the coefficient matrix. The corresponding eigenvectors contain the amplitudes of the basis functions that represent the fields on the conductor surfaces of the modes.

8.5 The incoming fields

8.5.1 *The spatial incoming fields*

In this section we calculate the spatial incoming fields (8.10) generated by an excitation wire i_j in layer i. In order not to obscure the notation we will omit the subscripts i_j and i. No confusion is possible because all quantities are related to a single wire. If we substitute the expansions for the longitudinal magnetic field and the radial derivative of the longitudinal electric field (8.9) into (8.5) we get the following expansions for the incoming fields:

$$E_x^{\text{in}}(\mathbf{r}) = \sum_{n=-\infty}^{+\infty} aA_n \int_0^{2\pi} G\, e^{jn\phi'}\, d\phi'$$

$$H_x^{\text{in}}(\mathbf{r}) = -\sum_{n=-\infty}^{+\infty} aB_n \int_0^{2\pi} \frac{\partial G}{\partial r'}\, e^{jn\phi'}\, d\phi'. \qquad (8.16)$$

a is the radius of the wire and $\mathbf{r}$ is located outside the wire, that is, $|\mathbf{r}| > a$. The integrals appearing in the right-hand sides of (8.16) can be integrated analytically by means of the following addition theorem for Bessel functions [11]:

$$H_0^{(2)}(\gamma|\mathbf{r}-\mathbf{r}'|) = \sum_{m=-\infty}^{+\infty} J_m(\gamma r') H_m^{(2)}(\gamma r)\, e^{jm(\phi-\phi')} \qquad (8.17)$$

with J_m (respectively $H_m^{(2)}$) is the Bessel function of order m (respectively the second-kind Hankel function of order m). Substitution of (8.17) in the Green's functions G of (8.16) allows us to perform the ϕ' integration. Only one term of the series in (8.17), namely the term with $m = n$, will not vanish. Hence, the integrals in (8.16) are given by

$$\int_0^{2\pi} G\, e^{jn\phi'}\, d\phi' = \frac{j\pi}{2} J_n(\gamma a) H_n^{(2)}(\gamma r)\, e^{jn\phi}$$

$$\int_0^{2\pi} \frac{\partial G}{\partial r'}\, e^{jn\phi'}\, d\phi' = \frac{j\pi}{2} J_n'(\gamma a) H_n^{(2)}(\gamma r)\, e^{jn\phi}. \qquad (8.18)$$

From the expansions (8.16) and the integrals (8.18) we get the expressions (8.10) with

$$C_n = \frac{ja\pi}{2} J_n(\gamma a) A_n$$

$$D_n = -\frac{ja\pi}{2} J_n'(\gamma a) B_n. \qquad (8.19)$$

8 LOSSLESS WIRE TRANSMISSION LINE STRUCTURES

Expressions (8.19) already yield the C_n and D_n coefficients in the eigensystem equations (8.15).

8.5.2 The spectral incoming fields

In this section the spectral incoming fields are calculated. These fields are needed to determine the scattered fields and to determine the contribution of the incoming fields of one wire to the total fields at the surface of another wire located in the same layer. In order to find the scattered fields we also have to decompose the spectral incoming fields into their TE and TM constituents.

8.5.2.1 The spectral longitudinal incoming fields

From (8.10) it is clear that the determination of the spectral incoming fields reduces to the calculation of the following Fourier transform:

$$F_n(k_y) = \frac{1}{2\pi} \int_{-\infty}^{+\infty} H_n^{(2)}(\gamma r) \, e^{j(n\phi + k_y y)} \, dy. \qquad (8.20)$$

First we take one step back and replace, by using (8.18), the Hankel function of order n by an angular integration over the Hankel function of order 0. The function $F_n(k_y)$ in (8.20) now becomes

$$F_n(k_y) = \frac{1}{(2\pi)^2 J_n(\gamma a)} \int_{-\infty}^{+\infty} e^{jk_y y} \int_0^{2\pi} H_0^{(2)}(\gamma |\mathbf{r}' - \mathbf{r}|) \, e^{jn\phi'} \, d\phi' \, dy. \qquad (8.21)$$

Interchanging the integration order yields two integrations that can be performed analytically. The first is the Fourier transformation of the Hankel function which is found in [12]:

$$\int_{-\infty}^{+\infty} H_0^{(2)}(\gamma |\mathbf{r}' - \mathbf{r}|) \, e^{jk_y y} \, dy = 2j \, \frac{e^{-\Gamma |z - z'|} \, e^{jk_y y'}}{\Gamma} \qquad (8.22)$$

with

$$\Gamma^2 = k_y^2 - \gamma^2. \qquad (8.23)$$

To obtain $F_n(k_y)$ only the angular integration over ϕ' remains to be performed:

$$F_n(k_y) = 2j \, \frac{1}{(2\pi)^2 J_n(\gamma a) \Gamma} \int_0^{2\pi} e^{jn\phi'} e^{jk_y y' - \Gamma |z - z'|} \, d\phi'. \qquad (8.24)$$

Equation (8.24) can be integrated analytically only if $(z - z')$ has the same sign over the whole integration interval or, in other words, on the whole

8.5 THE INCOMING FIELDS

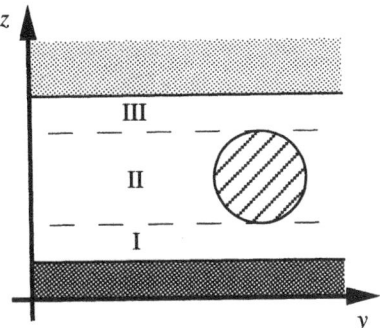

Fig. 8.4. Three regions in the excitation layer. Under the excitation wire (region I), next to the excitation wire (region II), and above the excitation wire (region III).

wire surface. This means that (8.24) can be calculated only for z-coordinates above and under the wire. In Fig. 8.4 this corresponds to regions I or III. In region II the integrand in (8.24) has to be divided into two parts and it is not possible to obtain a closed form for $F_n(k_y)$. Fortunately, in order to calculate the scattered fields we only need (8.24) at the layer interfaces of the layer in which the wire is embedded. So if we avoid wires which cross layer interfaces, as explicitly assumed at the beginning of this chapter, we can evaluate (8.24) analytically. Due to the change of sign in the exponential function we have to make a distinction between regions I and III. Let us start with region III. Equation (8.24) can now be written as

$$F_n^+(k_y) = 2\mathrm{j}\, \frac{\mathrm{e}^{-\Gamma(z-z_0)}\,\mathrm{e}^{\mathrm{j}k_y y_0}}{(2\pi)^2 J_n(\gamma a)\Gamma} \int_0^{2\pi} \mathrm{e}^{\mathrm{j}n\phi'}\,\mathrm{e}^{\mathrm{j}k_y(y'-y_0)+\Gamma(z'-z_0)}\,\mathrm{d}\phi'. \qquad (8.25)$$

The superscript '+' indicates that we are in region III above the wire. We have introduced the coordinates (y_0, z_0) of the centre of the wire. In the polar coordinate system of Fig. 8.2 the following relations hold:

$$\begin{aligned} y' - y_0 &= a \cos \phi' \\ z' - z_0 &= a \sin \phi'. \end{aligned} \qquad (8.26)$$

With this the second factor in the integrand of (8.25) becomes

$$\begin{aligned} \mathrm{e}^{\mathrm{j}k_y a \cos \phi' + \Gamma a \sin \phi'} &= \mathrm{e}^{\mathrm{j}a\gamma(\sin\alpha \cos\phi' + \cos\alpha \sin\phi')} \\ &= \mathrm{e}^{\mathrm{j}a\gamma \sin(\alpha+\phi')} \end{aligned} \qquad (8.27)$$

with α defined as

8 LOSSLESS WIRE TRANSMISSION LINE STRUCTURES

$$\sin \alpha = \frac{k_y}{\sqrt{k_y^2 - \Gamma^2}}$$

$$\cos \alpha = -\frac{j\Gamma}{\sqrt{k_y^2 - \Gamma^2}}.$$

(8.28)

We now use the following generating function for the Bessel functions [12]:

$$e^{jz \sin \theta} = \sum_{m=-\infty}^{+\infty} e^{jm\theta} J_m(z).$$

(8.29)

The above expression allows us to expand (8.27) in a Bessel series:

$$e^{ja\gamma \sin(\alpha + \phi')} = \sum_{m=-\infty}^{+\infty} e^{jm(\alpha + \phi')} J_m(a\gamma).$$

(8.30)

We substitute this back into (8.25) and we interchange the summation and integration. After integration, only the term corresponding to $m = -n$ does not vanish and (8.25) finally becomes

$$F_n^+(k_y) = 2j \frac{e^{-\Gamma(z-z_0)} e^{jk_y y_0}}{2\pi J_n(\gamma a)\Gamma} J_{-n}(\gamma a) e^{-jn\alpha}$$

(8.31)

or, after some manipulation, the final result for (8.20) is found to be

$$F_n^+(k_y) = j \frac{(-1)^n e^{-\Gamma(z-z_0)} e^{jk_y y_0}}{\pi \Gamma} \left(\frac{\Gamma + k_y}{\Gamma - k_y}\right)^{n/2}.$$

(8.32)

This is valid for $z > z_0 + a$. In region I we obtain in an analogous way the following expression for (8.20):

$$F_n^-(k_y) = j \frac{e^{\Gamma(z-z_0)} e^{jk_y y_0}}{\pi \Gamma} \left(\frac{\Gamma - k_y}{\Gamma + k_y}\right)^{n/2}$$

(8.33)

valid for $z < z_0 - a$ as indicated by the superscript '−'.

In both expressions (8.32) and (8.33) an exponential decaying wave appears. Equation (8.32) corresponds to an up-propagating incoming wave travelling in the positive z-direction. Equation (8.33) on the other hand is a down-propagating incoming wave travelling in the negative z-direction. With (8.32), the Fourier transform of the incoming fields (8.10) above the wire becomes

$$E_x^{in,+}(k_y, z) = \frac{e^{-\Gamma(z-z_0) + jk_y y_0}}{\Gamma} \sum_{n=-\infty}^{+\infty} (-1)^n \frac{j}{\pi} C_n \left(\frac{\Gamma + k_y}{\Gamma - k_y}\right)^{n/2}$$

$$H_x^{in,+}(k_y, z) = \frac{e^{-\Gamma(z-z_0) + jk_y y_0}}{\Gamma} \sum_{n=-\infty}^{+\infty} (-1)^n \frac{j}{\pi} D_n \left(\frac{\Gamma + k_y}{\Gamma - k_y}\right)^{n/2},$$

(8.34)

8.5 THE INCOMING FIELDS

and under the wire the following expressions are valid:

$$E_x^{\text{in},-}(k_y, z) = \frac{e^{\Gamma(z-z_0)+jk_y y_0}}{\Gamma} \sum_{n=-\infty}^{+\infty} \frac{j}{\pi} C_n \left(\frac{\Gamma - k_y}{\Gamma + k_y}\right)^{n/2}$$

$$H_x^{\text{in},-}(k_y, z) = \frac{e^{\Gamma(z-z_0)+jk_y y_0}}{\Gamma} \sum_{n=-\infty}^{+\infty} \frac{j}{\pi} D_n \left(\frac{\Gamma - k_y}{\Gamma + k_y}\right)^{n/2}.$$

(8.35)

8.5.2.2 *The incoming fields as sources in the TE and TM cascades* So far we have calculated only the spectral longitudinal incoming fields. In this section we determine the sources in the TE and TM cascades corresponding to these incoming fields. Once these sources are known we can find the spectral scattered fields in every layer by solving the cascades with the techniques described in Chapter 6.

We first rewrite the incoming fields (8.34) and (8.35) in a more suitable form:

$$E_x^{\text{in},+}(k_y, z) = e^+(k_y) e^{-\Gamma_e(z-z_{e,e+1})}$$
$$H_x^{\text{in},+}(k_y, z) = h^+(k_y) e^{-\Gamma_e(z-z_{e,e+1})}$$

(8.36)

and

$$E_x^{\text{in},-}(k_y, z) = e^-(k_y) e^{-\Gamma_e(z_{e-1,e}-z)}$$
$$H_x^{\text{in},-}(k_y, z) = h^-(k_y) e^{-\Gamma_e(z_{e-1,e}-z)}.$$

(8.37)

At this point in the calculation the subscript e, $z_{e,e+1}$ and $z_{e-1,e}$ are introduced to explicitly refer to the excitation layer, that is, the layer in which the considered source wire is located (Fig. 8.5). Expressions for $e^+(k_y)$ and $h^+(k_y)$ (respectively $e^-(k_y)$ and $h^-(k_y)$) can be easily found by identifying (8.36) with (8.34) and (8.37) with (8.35).

First we will concentrate on the up-going incoming fields (8.36). These fields will be represented as a (V,I) source at the top of the excitation

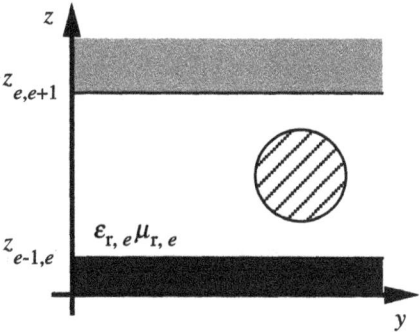

Fig. 8.5. An excitation layer with a source wire.

layer in the TE and TM cascades. We subdivide these incoming fields into their TE and TM parts. In Chapter 6 it is shown that the general form of an up-going TM wave is given by

$$E'^{\text{in}}(k_y, z) = A'^{\text{in}} e^{-\Gamma_e(z - z_{e, e+1})}$$
$$H''^{\text{in}}(k_y, z) = (A'^{\text{in}}/Z_{\text{TM}, e}) e^{-\Gamma_e(z - z_{e, e+1})} \tag{8.38}$$

with A'^{in} the amplitude of the wave. An up-going TE wave with amplitude A''^{in} is represented by

$$E''^{\text{in}}(k_y, z) = A''^{\text{in}} e^{-\Gamma_e(z - z_{e, e+1})}$$
$$H'^{\text{in}}(k_y, z) = (A''^{\text{in}}/Z_{\text{TE}, e}) e^{-\Gamma_e(z - z_{e, e+1})}. \tag{8.39}$$

Combining (8.38) and (8.39) using equation (6.17) yields expressions for the longitudinal incoming fields:

$$E_x^{\text{in}}(k_y, z) = \frac{\beta A'^{\text{in}} - k_y A''^{\text{in}}}{k^2} e^{-\Gamma_e(z - z_{e, e+1})}$$
$$H_x^{\text{in}}(k_y, z) = \frac{\beta A''^{\text{in}}/Z_{\text{TE}, e} - k_y A'^{\text{in}}/Z_{\text{TM}, e}}{k^2} e^{-\Gamma_e(z - z_{e, e+1})} \tag{8.40}$$

with $k^2 = k_y^2 + \beta^2$. Comparing (8.40) and (8.36) leads to two equations in the unknown amplitudes A'^{in} and A''^{in}. If we solve these equations, we get

$$A'^{\text{in}} = \frac{k^2}{k_y^2 - \beta^2 \dfrac{Z_{\text{TM}, e}}{Z_{\text{TE}, e}}} \left(-\beta \frac{Z_{\text{TM}, e}}{Z_{\text{TE}, e}} e^+ - k_y Z_{\text{TM}, e} h^+ \right)$$

$$A''^{\text{in}} = \frac{k^2}{k_y^2 - \beta^2 \dfrac{Z_{\text{TM}, e}}{Z_{\text{TE}, e}}} (-k_y e^+ - \beta Z_{\text{TM}, e} h^+). \tag{8.41}$$

Equation (8.38) is represented in the TM cascade as a (V, I) source equal to $(A'^{\text{in}}, A'^{\text{in}}/Z_{\text{TM}, e})$ located at the top $z_{e, e+1}$ of the excitation layer. Equation (8.39) is a (V, I) source in the TE cascade equal to $(A''^{\text{in}}, A''^{\text{in}}/Z_{\text{TE}, e})$ and also located at $z_{e, e+1}$.

The down-going incoming fields (8.37) are represented by a (V, I) source at the bottom of the excitation layer. The reader can verify that these fields are represented as a TM source equal to $(-B'^{\text{in}}, B'^{\text{in}}/Z_{\text{TM}, e})$ and a TE source equal to $(-B''^{\text{in}}, B''^{\text{in}}/Z_{\text{TE}, e})$ at the bottom $z_{e-1, e}$ of the excitation layer. B'^{in}

and B''^{in} are equal to

$$B'^{in} = \frac{k^2}{k_y^2 - \beta^2 \dfrac{Z_{TM,e}}{Z_{TE,e}}} \left(-\beta \frac{Z_{TM,e}}{Z_{TE,e}} e^- + k_y Z_{TM,e} h^- \right)$$

$$B''^{in} = \frac{k^2}{k_y^2 - \beta^2 \dfrac{Z_{TM,e}}{Z_{TE,e}}} (-k_y e^- + \beta Z_{TM,e} h^-)$$

(8.42)

8.5.3 Angular decomposition of the incoming fields on the wire surfaces

8.5.3.1 On the excitation wire
The derivation in Section 8.5.1 resulted in the angular representation (8.9) of the spatial incoming fields. From this representation we immediately find the required decomposition (8.10) at the wire surface by taking r_{i_j} equal to a_{i_j}.

8.5.3.2 On an observation wire in the excitation layer
In this section we consider an excitation wire, denoted with the subscript 'e', and an observation wire, denoted with the subscript 'o'. Both wires are in the same layer and we simply omit subscripts referring to this layer.

To find the angular Fourier decomposition of the incoming fields of the excitation wire at the surface of the observation wire, the incoming fields are first spatially Fourier transformed and calculated at the observation wire. In a second step these fields are inverse Fourier transformed and decomposed into a Fourier series valid on the surface of the observation wire.

To bring the above programme to a satisfactory result, it must be remembered that it was impossible to analytically obtain the Fourier transform of the incoming fields in region II of Fig. 8.4 (see Section 8.5.2). This is the case when the observation wire is (partially) located in this region. To circumvent this difficulty a new cartesian coordinate system (Fig. 8.6) $(x, \tilde{y}, \tilde{z})$ is introduced with its origin at the excitation wire and with the positive $\tilde{z}$-axes pointing to the centre of the observation wire. $\Phi_{e,o}$ represents the rotation angle between the (x, y, z) and $(x, \tilde{y}, \tilde{z})$ coordinate systems. $d_{e,o}$ is the distance between both wires. Two new polar coordinate systems $(r_e, \tilde{\phi}_e, x)$ and $(r_o, \tilde{\phi}_o, x)$ relative to the new cartesian coordinate system are used. The relations with the old polar coordinate systems associated with each wire (see Fig. 8.2) are given by

$$\tilde{\phi}_e = \phi_e - \Phi_{e,o} \qquad \tilde{\phi}_o = \phi_o - \Phi_{e,o}.$$

(8.43)

The incoming electric field (8.10) from wire e in the new polar coordinate

186   8 LOSSLESS WIRE TRANSMISSION LINE STRUCTURES

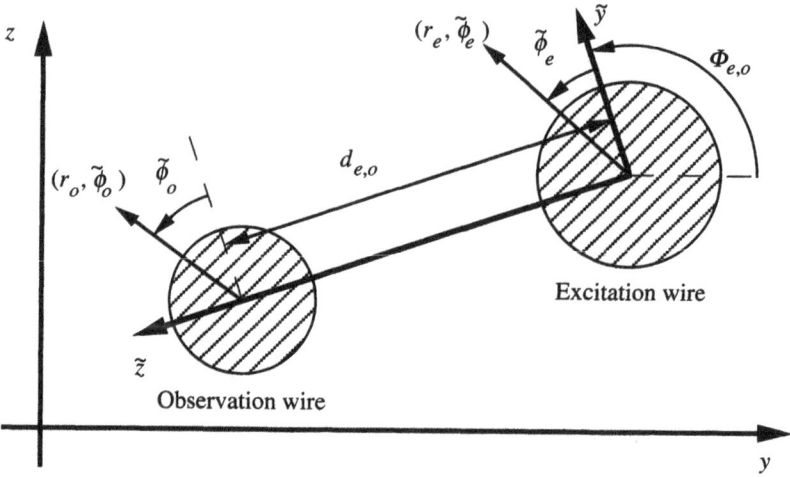

Fig. 8.6. New cartesian coordinate system with corresponding local polar coordinate systems.

system is equal to

$$E^{in}_{x,e}(\mathbf{r}) = \sum_{n=-\infty}^{+\infty} C_{n,e} e^{jn\Phi_{e,o}} H_n^{(2)}(\gamma_e r_e) e^{jn\tilde{\phi}_e}. \tag{8.44}$$

Now this incoming field is spatially Fourier transformed in the $\tilde{y}$-direction. We use no new spectral variable but keep to the notation k_y. The calculation of the Fourier transform is identical to the spectral transformation in Section 8.5.2. Hence

$$E^{in}_{x,e}(k_y, \tilde{z}) = \frac{e^{\Gamma_e \tilde{z}}}{\Gamma_e} \sum_{n=-\infty}^{+\infty} \frac{j}{\pi} (-1)^n C_{n,e} \left(\frac{\Gamma_e + k_y}{\Gamma_e - k_y}\right)^{n/2} e^{jn\Phi_{e,o}}. \tag{8.45}$$

The last step in the calculation of the Fourier coefficients at the observation wire consists of the inverse Fourier transformation of the field (8.45) with respect to k_y and in determining the Fourier series expansion at the surface of the observation wire. This means that we are looking for the Fourier coefficients in the expansions (8.12) and (8.13). In the simplified notation of this section these coefficients are denoted by $E^o_{n,e}$ and $F^o_{n,e}$. For $E^o_{n,e}$ we find

$$E^o_{n,e} = \int_{-\infty}^{+\infty} \left[\frac{e^{-jn\Phi_{e,o}}}{\Gamma_e} \sum_{m=-\infty}^{+\infty} \frac{j}{\pi} (-1)^m C_{m,e} \left(\frac{\Gamma_e + k_y}{\Gamma_e - k_y}\right)^{m/2} e^{jm\Phi_{e,o}} e^{-\Gamma_e d_{e,o}} \right. \tag{8.46}$$
$$\left. \times \int_0^{2\pi} e^{-jn\tilde{\phi}_o} e^{-\Gamma_e(\tilde{z}-d_{e,o})} e^{-jk_y y} d\tilde{\phi}_o \right] dk_y.$$

In (8.46) the inverse Fourier transformation is performed after the angular integration. The angular integration is determined analytically in the same

8.6 THE SCATTERED FIELDS

way as in (8.25). This finally leads us to

$$E^o_{n,e} = e^{-jn\Phi_{e,o}} J_n(\gamma_e a_o) \sum_{m=-\infty}^{+\infty} \frac{j}{\pi}(-1)^m C_{m,e} e^{jm\Phi_{e,o}}$$
$$\int_{-\infty}^{+\infty} \frac{e^{-\Gamma_e d_{e,o}}}{\Gamma_e} \left(\frac{\Gamma_e + k_y}{\Gamma_e - k_y}\right)^{(m-n)/2} dk_y. \quad (8.47)$$

The reader can verify that the calculation for the magnetic field results in an analogous expression for $F^o_{n,e}$:

$$F^o_{n,e} = e^{-jn\Phi_{e,o}} J_n(\gamma_e a_o) \sum_{m=-\infty}^{+\infty} \frac{j}{\pi}(-1)^m D_{m,e} e^{jm\Phi_{e,o}}$$
$$\int_{-\infty}^{+\infty} \frac{e^{-\Gamma_e d_{e,o}}}{\Gamma_e} \left(\frac{\Gamma_e + k_y}{\Gamma_e - k_y}\right)^{(m-n)/2} dk_y. \quad (8.48)$$

Expressions (8.47) and (8.48) yield the E_n and F_n coefficients in the eigensystem equations (8.15).

8.6 The scattered fields

8.6.1 *The spectral scattered fields*

In Section 8.5.2.2 the spectral incoming fields are represented as equivalent sources at the top and bottom of the excitation layer both for the TE and TM cascades of the layered medium. In this section we will determine the solution of these cascades in every layer. In other words we will determine the TE and TM components of the spectral scattered fields in every layer. Both the excitation wire and the excitation layer are denoted with the subscript e. Because the solution techniques for both the TM cascade and the TE cascade are identical, we will use the generic notation of Section 6.2.5. As shown in Section 8.5.2.2 the (V, I) source at the top (superscript '+') of the excitation layer is given by

$$\begin{bmatrix} V \\ I \end{bmatrix}^+ = \begin{bmatrix} A''^{\text{in}} \\ \dfrac{A''^{\text{in}}}{Z_e} \end{bmatrix}. \quad (8.49)$$

The second source at the bottom (superscript '−') of the excitation layer is equal to

$$\begin{bmatrix} V \\ I \end{bmatrix}^- = \begin{bmatrix} -B''^{\text{in}} \\ \dfrac{B''^{\text{in}}}{Z_e} \end{bmatrix}. \quad (8.50)$$

188   8 LOSSLESS WIRE TRANSMISSION LINE STRUCTURE

The relevant A and B values are given by (8.41) and (8.42). The total solution of the cascade due to both sources is equal to the superposition of the solutions due to both sources separately. The solution of the cascade for a general (V, I) source is presented in Chapter 6. The wave coefficients corresponding to the source (8.49) are denoted by R_1^+ and R_L^+, and those corresponding to the source (8.50) by R_1^- and R_L^-.

Three different situations have to be considered in the determination of the scattered fields. In the first situation we determine the scattered fields in a layer i under the excitation layer. The fields for an arbitrary z-coordinate in this layer can then be written as (see Chapter 6):

$$\mathbf{C}_i(z) = e^{-\Gamma_i(z_{i,i+1}-z)} \left(\prod_{k=e-1}^{i+1} e^{-\Gamma_k d_k} \right) \mathbf{N}_i^u(z) \left(\prod_{k=1}^{i-1} \mathbf{N}_k^u \right) \mathbf{K}_1 R_1^-$$
$$+ e^{-\Gamma_i(z_{i,i+1}-z)} \left(\prod_{k=e}^{i+1} e^{-\Gamma_k d_k} \right) \mathbf{N}_i^u(z) \left(\prod_{k=1}^{i-1} \mathbf{N}_k^u \right) \mathbf{K}_1 R_1^+ .$$

(8.51)

The second situation deals with a layer i above the excitation layer. The fields for an arbitrary z-coordinate in this layer can now be written as (see Chapter 6):

$$\mathbf{C}_i(z) = e^{-\Gamma_i(z-z_{i-1,i})} \left(\prod_{k=e}^{i+1} e^{-\Gamma_k d_k} \right) \mathbf{N}_i^d(z) \left(\prod_{k=L}^{i+1} \mathbf{N}_k^d \right) \mathbf{K}_L R_L^-$$
$$+ e^{-\Gamma_i(z-z_{i-1,i})} \left(\prod_{k=e+1}^{i-1} e^{-\Gamma_k d_k} \right) \mathbf{N}_i^d(z) \left(\prod_{k=L}^{i+1} \mathbf{N}_k^d \right) \mathbf{K}_L R_L^+ .$$

(8.52)

In the third and last situation the scattered fields in the excitation layer are determined. The fields for an arbitrary z-coordinate in the excitation layer turn out to be

$$\mathbf{C}_e(z) = e^{-\Gamma_e(z-z_{e-1,e})} \mathbf{N}_e^d(z) \left(\prod_{k=L}^{e+1} \mathbf{N}_k^d \right) \mathbf{K}_L R_L^-$$
$$+ e^{-\Gamma_e(z_{e,e+1}-z)} \mathbf{N}_e^u(z) \left(\prod_{k=1}^{e-1} \mathbf{N}_k^u \right) \mathbf{K}_1 R_1^+ .$$

(8.53)

8.6.2 *The spatial scattered fields and the angular decomposition on the wire surfaces*

In this section we determine the spatial scattered fields and the angular decomposition on the surface of an observation wire 'o' in observation layer 'o'. First we explicitly introduce the z-coordinate $z_{0,o}$ of the centre of the observation wire as the reference z-coordinate in expressions (8.51), (8.52), and (8.53). The reader can verify that these expressions can then be cast into

8.6 THE SCATTERED FIELDS

the following form for the TM cascade:

$$E_e^{'sc,o} = A_e^{'sc,o} e^{-\Gamma_o(z-z_{0,o})} + B_e^{'sc,o} e^{\Gamma_o(z-z_{0,o})}$$
$$H_e^{''sc,o} = (1/Z_{TM,o})(A_e^{'sc,o} e^{\Gamma_o(z-z_{0,o})} - B_e^{'sc,o} e^{\Gamma_o(z-z_{0,o})}) \quad (8.54)$$

and for the TE cascade:

$$E_e^{''sc,o} = A_e^{''sc,o} e^{\Gamma_o(z-z_{0,o})} + B_e^{''sc,o} e^{\Gamma_o(z-z_{0,o})}$$
$$H_e^{'sc,o} = (1/Z_{TE,o})(A_e^{''sc,o} e^{-\Gamma_o(z-z_{0,o})} - B_e^{''sc,o} e^{\Gamma_o(z-z_{0,o})}). \quad (8.55)$$

From these equations, using equation (6.17), we find for the longitudinal scattered fields:

$$E_{x,e}^{sc,o} = A_{e,e}^{sc,o} e^{-\Gamma_o(z-z_{0,o})} + B_{e,e}^{sc,o} e^{\Gamma_o(z-z_{0,o})} \quad (8.56)$$

with

$$A_{e,e}^{sc,o} = \frac{\beta A_e^{'sc,o} - k_y A_e^{''sc,o}}{k^2}$$

$$B_{e,e}^{sc,o} = \frac{\beta B_e^{'sc,o} - k_y B_e^{''sc,o}}{k^2} \quad (8.57)$$

and

$$H_{x,e}^{sc,o} = A_{h,e}^{sc,o} e^{-\Gamma_o(z-z_{0,o})} + B_{h,e}^{sc,o} e^{\Gamma_o(z-z_{0,o})} \quad (8.58)$$

with

$$A_{h,e}^{sc,o} = \frac{-k_y A_e^{'sc,o}/Z_{TM,o} + \beta A_e^{''sc,o}/Z_{TE,o}}{k^2}$$

$$B_{h,e}^{sc,o} = \frac{k_y B_e^{'sc,o}/Z_{TM,o} - \beta B_e^{''sc,o}/Z_{TE,o}}{k^2}. \quad (8.59)$$

In (8.57) and (8.59) $k^2 = k_y^2 + \beta^2$. The Fourier series expansion coefficients $G_{n,e}^o$ and $H_{n,e}^o$ on the surface of the observation wire are found by inverse Fourier transforming the fields (8.56) and (8.58), followed by a Fourier decomposition. This leads us to integrals analogous to (8.46) in which the angular Fourier decomposition is determined analytically. The reader can easily verify that the final result is

$$G_{n,e}^o = J_n(\gamma_p a_o) \int_{-\infty}^{+\infty} e^{-jk_y y_o} \left[A_{e,e}^{sc,o}(-1)^n \left(\frac{\Gamma_o - k_y}{\Gamma_o + k_y}\right)^{n/2} \right.$$
$$\left. + B_{e,e}^{sc,o} \left(\frac{\Gamma_o + k_y}{\Gamma_o - k_y}\right)^{n/2} \right] dk_y \quad (8.60)$$

8 LOSSLESS WIRE TRANSMISSION LINE STRUCTURES

and

$$H^\circ_{n,e} = J_n(\gamma_o a_o) \int_{-\infty}^{+\infty} e^{-jk_y y_o} \left[A^{sc,o}_{h,e}(-1)^n \left(\frac{\Gamma_o - k_y}{\Gamma_o + k_y}\right)^{n/2} \right.$$
$$\left. + B^{sc,o}_{h,e} \left(\frac{\Gamma_o + k_y}{\Gamma_o - k_y}\right)^{n/2} \right] dk_y. \quad (8.61)$$

Expressions (8.60) and (8.61) yield the G_n and H_n coefficients in the eigensystem equations (8.15).

8.7 The inverse Fourier transformation

The remaining integrals in the expressions (8.47) and (8.48) for E_n and F_n, and (8.60) and (8.61) for G_n and H_n are evaluated numerically. These integrals are of the following canonical form:

$$\int_{-\infty}^{+\infty} f(k_y) e^{-jk_y \Delta y} \left(\frac{\Gamma + k_y}{\Gamma - k_y}\right)^{m/2} dk_y \quad m = \ldots, -2, -1, 0, 1, 2, \ldots \quad (8.62)$$

where Δy is the difference in y-coordinate between the centres of the observation and excitation wires. The function $f(k_y)$ contains the wave propagation information from excitation to observation wire. For $k_y \in [-k_c, k_c]$ the integral in (8.62) is evaluated with simple Gaussian quadrature integration formulas along a suitable complex integration path (cf. Chapter 6).

The remaining part of the integration, that is, $k_y \in [k_c, +\infty[$ and $k_y \in]-\infty, -k_c]$, allows a special treatment and is called the asymptotic integration. In this asymptotic integration the value of k_c is taken high enough such that Γ can be approximated by $|k_y|$ and such that we can use the simplified asymptotic cascade of Section 6.2.7.2 for the determination of the scattered fields.

In this asymptotic cascade the function $f(k_y)$ will contain a sum of terms of the following form:

$$f(k_y) \propto h(k_y) e^{-\Delta z k_y} \quad \Delta z \geqslant 0. \quad (8.63)$$

We have introduced the '$\propto$'-sign meaning 'proportional to'. Each of these terms represents a wave going from the excitation wire to the observation wire. Δz is the total distance, in the z-direction, propagated by the wave. This distance also includes possible reflections at layer interfaces and can be longer than the geometrical difference in z-coordinates of the centres between the excitation and observation wires. The function $h(k_y)$ contains, among

8.7 THE INVERSE FOURIER TRANSFORMATION

other things, the reflection coefficients and transmission coefficients from the layer interfaces passed by the wave.

We have to be careful with the last factor in the integrand of (8.62). If $m = 0$ there is no problem because that factor is equal to 1. Let us first assume that $m > 0$ and that $k_y \in [k_c, +\infty[$; then we have

$$\lim_{\Gamma \to k_y} \left(\frac{\Gamma + k_y}{\Gamma - k_y}\right)^{m/2} = \lim_{\Gamma \to k_y} \left(\frac{\sqrt{k_y^2 - \gamma^2} + k_y}{\sqrt{k_y^2 - \gamma^2} - k_y}\right)^{m/2} = \left(\frac{-4k_y^2}{\gamma^2}\right)^{m/2} \propto k_y^m. \quad (8.64)$$

If $m > 0$ and $k_y \in \,]-\infty, k_c]$, then we have

$$\lim_{\Gamma \to -k_y} \left(\frac{\Gamma + k_y}{\Gamma - k_y}\right)^{m/2} \propto k_y^{-m}. \quad (8.65)$$

For large k_y values, contribution (8.65) is negligible as compared to (8.64). Hence if $m > 0$ we neglect the negative asymptotic integration interval. If $m < 0$ the positive integration interval is neglected. If we change the integration variable k_y to $-k_y$ in the case $m < 0$, all the remaining asymptotic integrations can be cast into the following canonical form:

$$\int_{k_c}^{+\infty} k_y^m h(k_y) \, e^{-ak_y} \, dk_y \qquad \text{Re}(a) \geq 0 \quad (8.66)$$

with $m = 0, 1, 2, \ldots$, and $a = j\Delta y + \Delta z$.

As we know, the propagation constant or eigenvalue β is found by an iterative process. Consequently, at each frequency the coefficients in the system (8.15) have to be determined for several values of β. However, it can be shown that $h(k_y)$ is a polynomial in the propagation constant β:

$$h(k_y) = h_0(k_y) + \beta h_1(k_y) + \beta^2 h_2(k_y) + \cdots + \beta^n h_n(k_y). \quad (8.67)$$

If we insert this in (8.66), we get the following β-independent integrations:

$$\int_{k_c}^{+\infty} k_y^m h_i(k_y) \, e^{-ak_y} \, dk_y \qquad i = 0, 1, \ldots, n \quad \text{Re}(a) \geq 0. \quad (8.68)$$

This means that these asymptotic integrations have to be performed only once for each frequency.

Finally we can rewrite (8.68) as

$$\frac{e^{ak_c}}{a} \sum_{j=0}^{m} C_m^j \frac{k_c^{m-j}}{a^j} \int_0^{+\infty} x^j h_i\left(\frac{x}{a} + k_c\right) e^{-x} \, dx \quad (8.69)$$

where

$$C_m^j = \frac{m!}{j!(m-j)!}. \quad (8.70)$$

8　LOSSLESS WIRE TRANSMISSION LINE STRUCTURES

The integral in (8.69) is integrated numerically with a Gauss–Laguerre quadrature formula.

8.8　The longitudinal currents

The total longitudinal currents on the wire surfaces are determined from the eigenvector of the discrete eigenvalue problem. We will again omit the subscripts i_j or i referring to a specific wire or layer. The relevant coefficients of the eigenvector are A_n and B_n. The total longitudinal current I is equal to the integral of the tangential or ϕ component of the total magnetic field at the wire surface:

$$I = a \int_0^{2\pi} H_\phi \, d\phi. \tag{8.71}$$

In Appendix A, we find from Maxwell's equations the following relation between the tangential magnetic field and the longitudinal fields:

$$H_\phi = -\frac{j\beta}{\gamma^2 a}\frac{\partial H_x}{\partial \phi} - \frac{j\omega\varepsilon}{\gamma^2}\frac{\partial E_x}{\partial r}. \tag{8.72}$$

Substituting the representations (8.9) in (8.72) gives us a Fourier series decomposition of the tangential magnetic field at the wire surface:

$$H_\phi = \sum_{n=-\infty}^{+\infty} \left[-\frac{j\beta}{\gamma^2 a}(jn)B_n - \frac{j\omega\varepsilon}{\gamma^2}A_n \right] e^{jn\phi}. \tag{8.73}$$

This expression is inserted in (8.71) and integrated analytically. Only the 'd.c. component' proportional to A_0 does not vanish after integration. Hence, the total longitudinal current is given by

$$I = -2\pi a \frac{j\omega\varepsilon}{\gamma^2} A_0. \tag{8.74}$$

8.9　The propagated power

8.9.1　Introduction

In this section we determine the power propagated by the modes of the structure and the cross-power propagated by two different modes. These powers together with the longitudinal currents are needed for the impedance matrix of the equivalent transmission line model as shown in Chapter 4.

8.9 THE PROPAGATED POWER

We make a distinction between layers with wires and layers without wires. In the latter case we directly apply the spectral domain technique described in Section 6.4. In the former case, calculations are more complicated and we will use a spatial domain technique as explained below.

8.9.2 Power propagated in a layer with wires

In a layer with wires or in other words in an excitation layer, the total fields consist of incoming and scattered parts. If, as in Section 6.4, we want to calculate the propagated power in the spectral domain, we need both parts. In order to perform the z integration analytically, we need explicit analytical expressions for these spectral fields. The scattered fields and the incoming fields in regions I and III of Fig. 8.4 are combinations of waves with an exponential z dependence. This is their only z dependence; see for example equations (8.34), (8.56), and (8.58). Hence, the integration can be performed analytically in a simple way. The incoming fields in region II of Fig. 8.4 however consist of travelling waves with z-dependent wave coefficients. As mentioned in Section 8.5.2.1 these wave coefficients cannot be obtained in closed form. This also means that the z integration in the power evaluation has to be performed numerically, which implies that the advantage of working in the spectral domain is lost. There are two possible ways to proceed.

First we can stay in the spectral domain and evaluate both the spectral integration over k_y and the z integration numerically. Note that the evaluation of the spectral incoming fields in region II is difficult.

Second we can evaluate the propagated power in the spatial domain by integrating the x-component of Poynting's vector over the cross-section of the excitation layer. For this we need the spatial y- and z-components of the electric and magnetic field. This is no problem for the spatial incoming fields, because we know these fields everywhere in the excitation layer. The spatial scattered fields however involve a numerical inverse Fourier transformation.

We opted for the second approach because it is easier to obtain the inverse Fourier transform of the scattered fields than to determine the spectral incoming fields in region II.

The power in the excitation layer is integrated numerically by introducing a grid as shown in Fig. 8.7. This grid consists of rectangular cells and special cells near the surface of the wires. The density increases in the neighbourhood of the wires where the fields change rapidly. At some distance away from the wires the grid is terminated in the y-direction. The boundaries of the grid are chosen such that the contribution of the region outside the grid to the power is negligible. The power is calculated as a sum over contributions of each cell. The fields in each cell, and hence the power, are taken to be

8 LOSSLESS WIRE TRANSMISSION LINE STRUCTURES

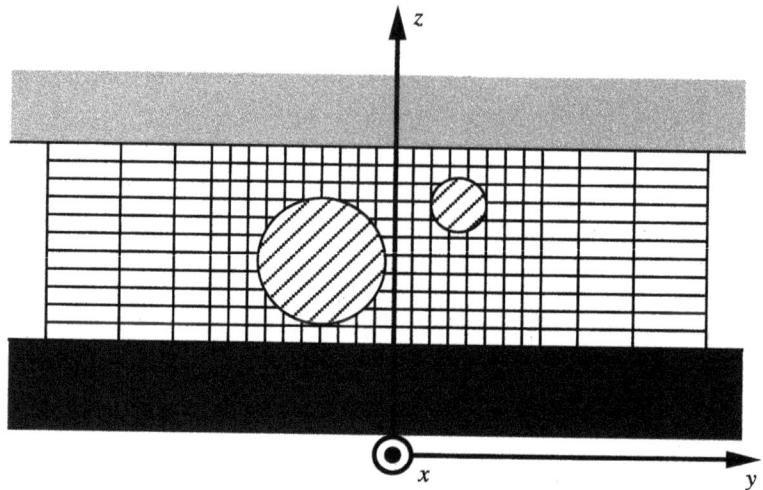

Fig. 8.7. Grid in an excitation layer used for the numerical integration of the propagated power.

constant. The power contribution of a cell is then given by the product of the area of the cell and the x-component of Poynting's vector at the centre of gravity of the cell.

8.9.2.1 *The spatial incoming fields* The y- and z-components can be calculated from the longitudinal field components with the following relations (cf. Appendix A):

$$E_y^{in} = -\frac{j\beta}{\gamma^2}\frac{\partial E_x^{in}}{\partial y} - \frac{j\omega\mu}{\gamma^2}\frac{\partial H_x^{in}}{\partial z}$$

$$E_z^{in} = -\frac{j\beta}{\gamma^2}\frac{\partial E_x^{in}}{\partial z} + \frac{j\omega\mu}{\gamma^2}\frac{\partial H_x^{in}}{\partial y}$$

(8.75)

and

$$H_y^{in} = -\frac{j\beta}{\gamma^2}\frac{\partial H_x^{in}}{\partial y} + \frac{j\omega\varepsilon}{\gamma^2}\frac{\partial E_x^{in}}{\partial z}$$

$$H_z^{in} = -\frac{j\beta}{\gamma^2}\frac{\partial H_x^{in}}{\partial z} - \frac{j\omega\varepsilon}{\gamma^2}\frac{\partial E_x^{in}}{\partial y}.$$

(8.76)

Again for notational simplicity all subscripts refering to layers and wires are omitted. The longitudinal incoming fields are given by the series expansions (8.10). In (8.75) and (8.76) we need the partial derivatives in the y- and z-direction of these expansions. The y derivative of the longitudinal electric field, for example, is given by

$$\frac{\partial E_x^{in}}{\partial y} = \sum_{n=-\infty}^{+\infty} C_n e^{jn\phi} \left[\cos \phi \gamma H_n'^{(2)}(\gamma a) - jn \frac{\sin \phi}{a} H_n^{(2)}(\gamma a) \right]. \quad (8.77)$$

The other derivatives are left to the reader.

8.9.2.2 *The spatial scattered fields* The y- and z-components of the scattered fields are easily found from the expressions for the TE and TM components (8.54) and (8.55). After combining the TE and TM components and inverse Fourier transformation, we find

$$E_y^{sc}(\mathbf{r}) = \int_{-\infty}^{+\infty} (1/k^2)[(A'^{sc}k_y + A''^{sc}\beta) e^{-\Gamma(z-z_0)}$$
$$+ (B'^{sc}k_y + B''^{sc}\beta) e^{\Gamma(z-z_0)}] e^{-jk_y y_0} dk_y \quad (8.78)$$

$$E_z^{sc}(\mathbf{r}) = \int_{-\infty}^{+\infty} (-j/\Gamma)[A'^{sc} e^{-\Gamma(z-z_0)} - B'^{sc} e^{\Gamma(z-z_0)}] e^{-jk_y y_0} dk_y \quad (8.79)$$

$$H_y^{sc}(\mathbf{r}) = \int_{-\infty}^{+\infty} (1/k^2)[(A'^{sc}k_y/Z_{TE} + A''^{sc}\beta/Z_{TM}) e^{-\Gamma(z-z_0)}$$
$$+ (B'^{sc}k_y/Z_{TE} + B''^{sc}\beta/Z_{TM}) e^{\Gamma(z-z_0)}] e^{-jk_y y_0} dk_y \quad (8.80)$$

$$H_z^{sc}(\mathbf{r}) = \int_{-\infty}^{+\infty} (1/\omega\mu)[A''^{sc} e^{-\Gamma(z-z_0)} + B''^{sc} e^{\Gamma(z-z_0)}] e^{-jk_y y_0} dk_y. \quad (8.81)$$

8.10 Case studies

8.10.1 *A single wire in a two-layered medium with ground plane*

The structure under consideration, consisting of a two-layered medium (Fig. 8.8) with a ground plane and a single wire, is characterized by four parameters: the relative permittivities $\varepsilon_{r,1}$ and $\varepsilon_{r,2}$ of substrate and top layer, the ratio of the height H of the centre of the wire to the thickness of the substrate d, and the ratio of the radius a of the wire to the thickness d. We assume that the layers are not magnetic; hence $\mu_{r,1} = \mu_{r,2} = 1$.

In our examples we took $a/d = 0.25$ and the relative permittivity of the substrate $\varepsilon_{r,1} = 4$. We assumed the top layer to be air with $\varepsilon_{r,2} = 1$. We used four different values of H/d: 0.5, 0.75, 1.25, and 1.5. The first two values (Fig. 8.8(b)) of H/d correspond to a wire in the substrate. This structure is analogous to Multiwire® or Microwire® configurations. The last two values (Fig. 8.8(a)) of H/d correspond to a wire in the air above the substrate. This

196   8 LOSSLESS WIRE TRANSMISSION LINE STRUCTURES

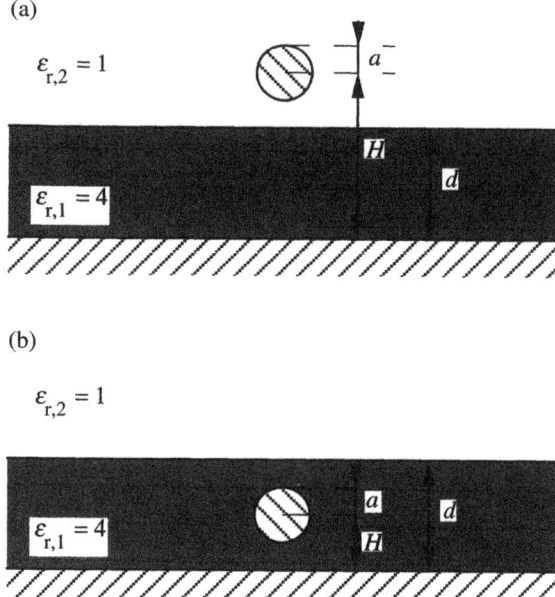

Fig. 8.8. A single wire in a microstrip configuration: (b) wire in the substrate; (a) wire in the air.

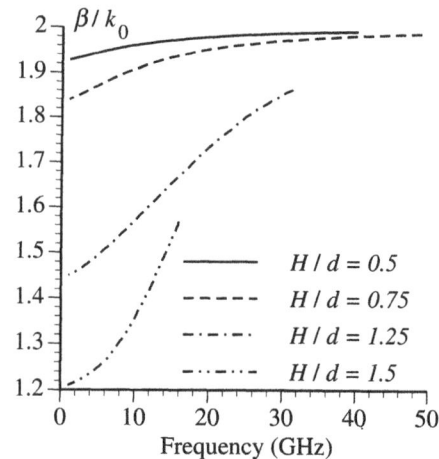

Fig. 8.9. Normalized propagation constant for the configuration of Fig. 8.8 for four different values of H/d.

models, for example, a bonding wire to a package in a high-speed integrated circuit. In Fig. 8.9 the normalized propagation constant β/k_0, and in Fig. 8.10 the impedance Z, are presented for the different values of H/d for a thickness d of the substrate equal to 3 mm.

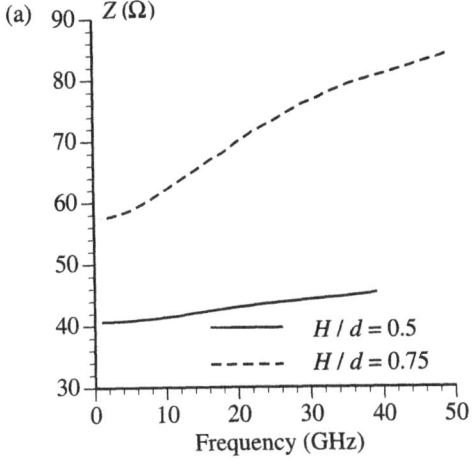

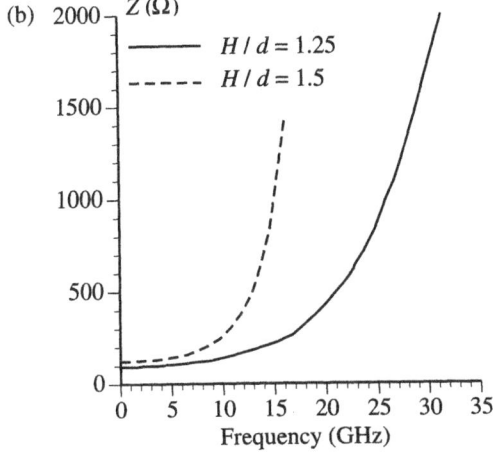

Fig. 8.10. Impedance for the configuration of Fig. 8.8: (a) wire in the substrate; (b) wire in the air.

For increasing frequencies, the fields become more and more confined in the substrate, due to dielectric guidance, resulting in an increasing effective permittivity (Fig. 8.9). At high frequencies and when the wire is in the substrate ($H/d = 0.5$ and $H/d = 0.75$), the normalized propagation constants almost reach their maximum possible value of 2. At these frequencies almost all the fields are confined inside the substrate. Figure 8.10 clearly illustrates the quasi-TEM behaviour of the impedance. The impedance remains almost constant when the propagation constant has already changed considerably. At 10 GHz the impedance suddenly increases dramatically, certainly for wires in the top layer. Moreover, the relative change of the propagation constant

8 LOSSLESS WIRE TRANSMISSION LINE STRUCTURES

is high for wires in the air. This is due to a strong deformation of the fields when the frequency increases.

At low frequencies the structure behaves quasi-statically; the longitudinal field components are negligible. The fields will be almost equal to the fields in a semi-infinite homogeneous medium with relative permittivity $\varepsilon_{r,\text{eff}} = (\beta/k_0)^2$. The impedance of a wire in homogeneous space above a ground plane is calculated analytically in [13]:

$$Z = \sqrt{\frac{\mu}{\varepsilon}} \frac{1}{\pi} \operatorname{arcch}\left(\frac{H}{a}\right). \tag{8.82}$$

This formula with $\mu = \mu_0$ and $\varepsilon = \varepsilon_0 \varepsilon_{r,\text{eff}}$ can be used as verification for the propagation constant and impedance. We found a deviation between our results and (8.82) smaller than 0.5 per cent.

8.10.2 Two wires in a double-layered medium with ground plane

We selected the same medium as in the previous section but now with a pair of wires as depicted in Fig. 8.11. The wires have a radius $a = 0.75$ mm; the substrate has a thickness $d = 3$ mm and a relative permittivity $\varepsilon_{r,1} = 4$. The distance between the centres of the wires is denoted by s and the centres of the wires are located at a height $H = 2.25$ mm above the ground plane.

Figure 8.12 shows the frequency dependence of the propagation constants and impedances for the two fundamental modes of the structure when the distance $s = 2.5$ mm. Due to symmetry these are the even and odd modes of the structure. The characteristic impedances of each wire for both modes are equal due to the symmetry (even mode) or antisymmetry (odd mode) in the field pattern. In the notation of Chapter 4 we have that $Z_{\text{odd}} = Z_{11} = Z_{21}$ and $Z_{\text{even}} = Z_{12} = Z_{22}$. The even mode has a higher propagation constant because for that mode the fields are located more in the air as compared to the case of the odd mode. This is due to attraction of the opposite currents on both wires. The fields of the even mode repel each other, hence the fields are pushed into the substrate.

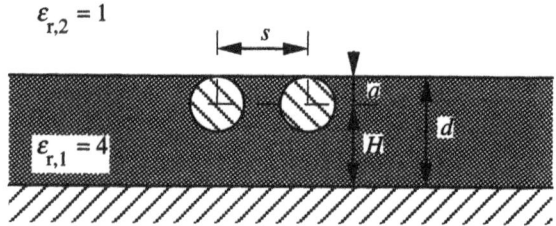

Fig. 8.11. Two wires in a microstrip configuration.

8.10 CASE STUDIES

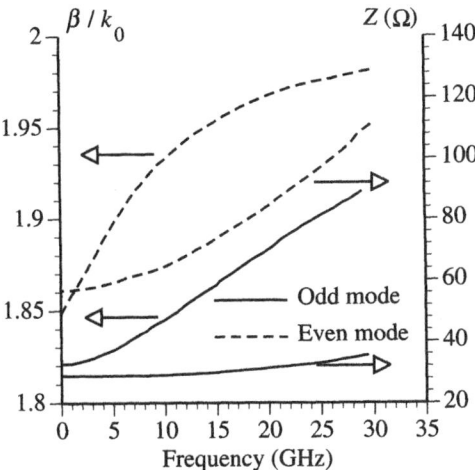

Fig. 8.12. Normalized propagation constants and impedances for the odd and even fundamental mode of the configuration in Fig. 8.11 ($s = 2.5$ mm).

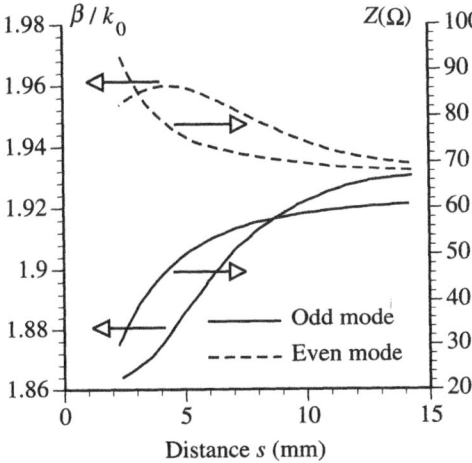

Fig. 8.13. Normalized propagation constants and impedances of the fundamental modes as a function of the distance between the wires of Fig. 8.11 at a frequency of 15 GHz.

In Fig. 8.13 the propagation constants and impedances are shown as a function of the distance s for a frequency equal to 15 GHz. For large separations of the wire the propagation constants of both modes evolve to the propagation constant of a single wire due to the decreased coupling between the wires. It is remarkable that the propagation constant of the even mode decreases, at some critical point, when the wires get closer. Because of the small distance the repelling of the fields is so high that the fields are now

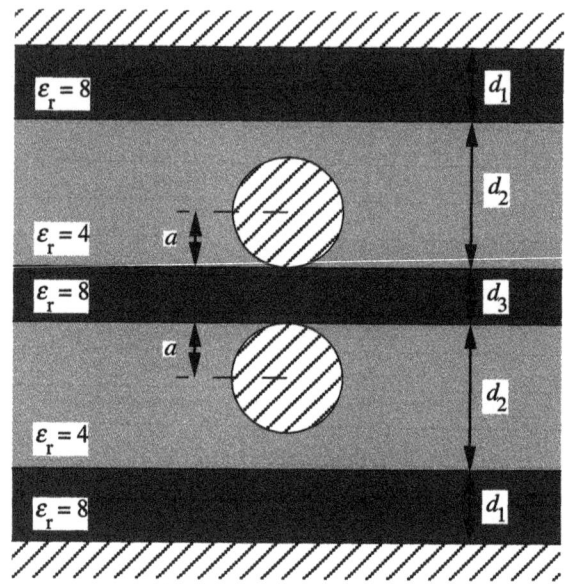

Fig. 8.14. Two wires in a multilayered stripline configuration.

pushed into the air above the wire zone. The characteristic impedance decreases for the even and increases for the odd mode when the wires are more separated.

8.10.3 *Two wires in a multilayered medium*

This last case study, although not directly related to a specific technological application, is deliberately selected to illustrate the capability of our theory and its numerical implementation to handle complex stratified media. The geometry of the five-layered medium is shown in Fig. 8.14. The various layer thicknesses are: $d_1 = 1$ mm, $d_2 = 2$ mm, and $d_3 = 0.75$ mm. The two wires have an identical radius $a = 0.75$ mm. The wires are located in the second and fourth layer both touching the third layer. Due to symmetry again an even and an odd mode is propagated. Figure 8.15 shows the circuit characteristics as a function of frequency.

The even mode has the lowest propagation constant because only a small part of the fields is in the third layer due to repulsion between the fields of both wires. The opposite occurs with the odd mode; the fields attract each other through the third layer. The higher permittivity of this third layer explains the higher effective permittivity or propagation constant. The dispersion as a function of frequency is limited for both modes because the

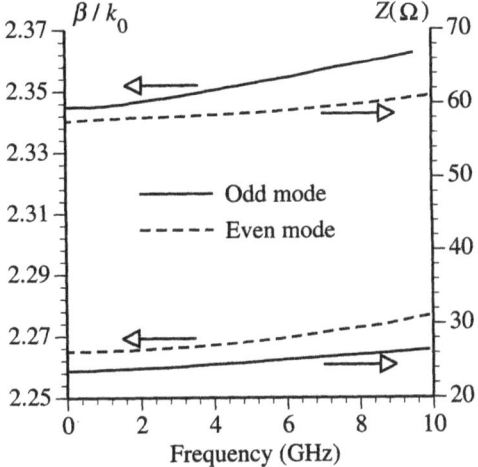

Fig. 8.15. Normalized propagation constant and impedances of the fundamental modes of the configuration in Fig. 8.14.

change in field pattern, as compared to the quasi-static situation for low frequencies, is limited.

References

[1] Shibata, H. and Terakado, R. (1984). Characteristics of transmission lines with a single wire for a multilayered circuit board. *IEEE Transactions on Microwave Theory and Techniques*, **MTT-32**, 4, 360–364.

[2] Wei, C., Harrington, R. F., Mautz, J. R., and Sarkar, T. K. (1984). Multiconductor transmission lines in multilayered dielectric media. *IEEE Transactions on Microwave Theory and Techniques*, **MTT-32**, 4, 439–450.

[3] Delbare, W. and De Zutter, D. (1989). Space-domain Green's function approach to the capacitance calculation of multiconductor lines in multilayered dielectrics with improved surface charge modeling. *IEEE Transactions on Microwave Theory and Techniques*, **MTT-37**, 10, 1562–1568.

[4] Olyslager, F., Faché, N., and De Zutter, D. (1991). New fast and accurate line parameter calculation of general multiconductor transmission lines in multilayered media. *IEEE Transactions on Microwave Theory and Techniques*, **MTT-39**, 6, 901–909.

[5] Faché, N. and De Zutter, D. (1989). Full wave analysis of a perfectly conducting wire transmission line in a double layered conductor backed medium. *IEEE Transactions on Microwave Theory and Techniques*, **MTT-37**, 3, 512–518.

[6] Faché, N., De Zutter, D., Olyslager, F., Van Hauwermeiren, L., Kok, P., and Botte, M. (1989). Theoretical and experimental modelling of discrete wire interconnection structures. In *Proceedings of the IEEE CHMT Third Conference on Electronic Packaging and Interconnections*, 119–124, Florence, Italy.

[7] Faché, N., Olyslager, F., and De Zutter, D. (1990). Eigenmodes of coupled perfectly conducting cylindrical wires in a multilayered dielectric medium. In *1990 International Symposium Digest IEEE Antennas and Propagation*, Vol. I, pp. 324–327, Dallas, Texas.

[8] Faché, N., Olyslager, F., and De Zutter, D. (1991). Full-wave analysis of coupled perfectly conducting wires in a multilayered medium. *IEEE Transactions on Microwave Theory and Techniques*, **MTT-39**, 4, 673–681.

[9] Michalski, K. A. and Zheng, D. (1989). Rigorous analysis of open microstrip lines of arbitrary cross-section in bound and leaky regimes. In *1989 IEEE MTT-S International Symposium Digest*, Vol. II, pp. 787–790, Long Beach, California.

[10] Michalski, K. A. and Zheng, D. (1989). Rigorous analysis of open microstrip lines of arbitrary cross-section in bound and leaky regimes. *IEEE Transactions on Microwave Theory and Techniques*, **MTT-37**, 12, 2005–2010.

[11] Magnus, W., Oberhettinger, F., and Soni, R. P. (1966). *Formulas and Theorems for the Special Functions of Mathematical Physics*, (3rd edn). Springer-Verlag, Berlin.

[12] Abramowitz, M. and Stegun, I. A. (1964). *Handbook of Mathematical Functions with Formulas, Graphs and Mathematical Tables*. Dover, New York.

[13] Collin, R. E. (1960). *Field Theory of Waves*, McGraw-Hill, New York.

9

ARBITRARILY-SHAPED POLYGONAL TRANSMISSION LINE STRUCTURES WITH FINITE CONDUCTIVITY

9.1 Introduction

As a third and most general transmission line structure, we handle transmission lines consisting of conductors with polygonal cross-section and finite conductivity. This kind of structure can for example be used for studying the influence of the thickness of strip transmission lines or to analyse transmission lines with trapezoidal cross-section due to under-etching. A conductor with a regular polygonal cross-section can serve as a model for a wire. The theory developed below also enables us to study the skin effect.

The waveguide structures analysed in this chapter consist of polygonal inhomogeneities in a multilayered medium. Though the theory elaborated in this chapter is valid for situations ranging from pure dielectric to perfectly conducting inhomogeneities, we will restrict ourselves to conductors. These conductors can be either perfectly conducting or have a finite conductivity.

Polygonal transmission lines have already been studied by a number of authors, but mostly in the quasi-TEM limit. In [1] a quasi-TEM analyses is presented of coupled polygonal conductors in a layered medium. The authors use the Green's function of free space and the layered medium is taken into account by polarization charges at the layer interfaces. In [2] the same authors used a perturbation theory to include losses in layers and conductors. In [3] and [4] the same kind of structures were analysed in the quasi-TEM limit, by means of the Green's function of the layered medium. Recently in [5] a complex image technique was proposed to handle the layered medium.

In [6] and [7] the rigorous full-wave analysis of open polygonal conductors is limited to the case of perfect conductors. The analysis is based on a mixed potential integral equation technique with the Green's dyadic of the layered medium. The integral equation is solved by the method of Galerkin. In this analysis the authors studied the behaviour of the propagation constants of the first- and higher-order modes in the bound and leaky regime. There exists a very abundant literature on the full-wave analyses of dielectric polygonal waveguides. We simply want to draw the reader's attention to two recent papers using an integral equation technique. In [8] a domain integral equation technique is presented. In that case the dielectric waveguides are replaced by equivalent contrast currents. A boundary integral equation technique for multilayered dielectric waveguides was presented in [9].

In this chapter the full circuit equivalent is presented including propagation constants and impedances for coupled polygonal transmission lines. The effect of losses in the conductors is investigated without any perturbations. Two of the present authors were the first to handle this full-wave analysis of lossless and lossy polygonal transmission lines including the calculation of the impedance. Publications concerning this work can be found in [10], [11], and [12]. This chapter is mainly based on the last publication [12].

The general outline of the theory is presented for conductors with arbitrary cross-section. In the detailed calculations, however, it becomes necessary to assume polygonal cross-sections. This assumption leads to considerable simplifications as a number of integrations can be carried out analytically.

The modes propagated along these structures are determined with a boundary integral equation technique. This integral equation is an extension of the integral equation used in the previous chapter for conductors with infinite conductivity. The sources of the modes are the tangential electric and magnetic field components at the conductor boundaries. The integral equation is solved by the method of Galerkin. The eigenvalues of this integral equation are the propagation constants, and the eigenvectors yield the modal fields. These modal fields are used to calculate the propagated power from which we determine the impedance matrix.

First a description of the geometry is given followed by the construction of the integral equation. Then this integral equation is discretized and a detailed calculation of the matrix coefficients of the discrete eigenvalue problem is given. Finally the impedance matrix is determined from the longitudinal currents and the propagated powers. In the last section of this chapter some case studies illustrate the theory.

9.2 Geometry

The geometry of the cross-section of the waveguide structure analysed in this chapter is depicted in Fig. 9.1. An arbitrary number of conductors with arbitrary cross-section are embedded in the general multilayered structure.

There are C conductors and we will use the subscript j ($j = 1, \ldots, C$) to refer to the jth conductor. Because a conductor can be located in more than one layer we use the notation C_i to indicate the total number of conductor parts in layer i while the subscript i_j ($j = 1, \ldots C_i$) refers to the jth conductor part in layer i.

9.2 GEOMETRY

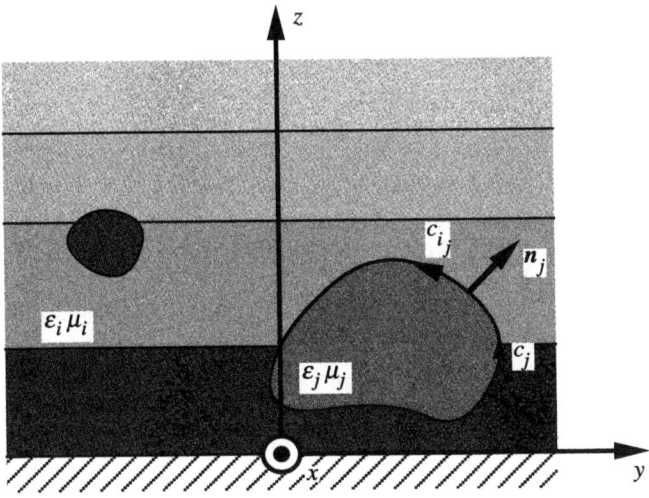

Fig. 9.1. Geometry of an arbitrarily-shaped transmission line structure.

Each conductor j consists of homogeneous and isotropic material characterized by an arbitrary permittivity ε_j and permeability μ_j. The permittivity is a complex number and contains the conductivity σ of the conductor. In the cases of interest for this book we will mostly deal with good conductors so that the real part of the relative permittivity ε_r can be neglected and $\varepsilon_r = \sigma/j\omega\varepsilon_0$. The case of perfectly conducting conductors is found by making ε_r infinite. The boundary of conductor j is denoted by c_j. Note that c_{i_j} is the boundary of conductor part j in layer i (see Fig. 9.1). $\mathbf{n}_j$ is the external normal to c_j and $\mathbf{r}_j$ is the position vector in the (y, z) plane of a point on c_j.

If a conductor part is located in a semi-infinite layer a fictitious layer boundary is introduced in order to make sure that all conductor parts are in a finite layer.

In the elaboration of the theory, and in the examples, we will not consider conductors consisting of multiple homogeneous regions as shown in Fig. 9.2. This is done only for notational simplicity. The whole theory can easily be extended to these configurations.

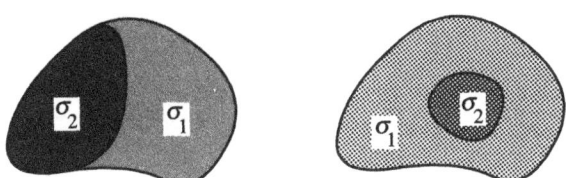

Fig. 9.2. Conductors consisting of multiple homogeneous regions.

9.3 Construction of the integral equation

9.3.1 Introduction

As in the previous chapters the common time and longitudinal dependence $e^{j(\omega t - \beta x)}$ in all field components is omitted. We use the longitudinal field components $E_x(\mathbf{r})$ and $H_x(\mathbf{r})$, with $\mathbf{r} = y\mathbf{1}_y + z\mathbf{1}_z$, as basis functions from which the other field components can be generated as shown in Appendix A. The sources which build up these longitudinal field components are the tangential fields $\mathbf{n}_j \times \mathbf{E}(\mathbf{r}_j)$ and $\mathbf{n}_j \times \mathbf{H}(\mathbf{r}_j)$ at the boundaries of the conductors. These fields consist of longitudinal contributions $E_{x,j}(\mathbf{r}_j)$ and $H_{x,j}(\mathbf{r}_j)$ along the conductors and transversal contributions $E_{t,j}(\mathbf{r}_j)$ and $H_{t,j}(\mathbf{r}_j)$ in the cross-section of the structure. The four functions $E_{x,j}(\mathbf{r}_j)$, $H_{x,j}(\mathbf{r}_j)$, $E_{t,j}(\mathbf{r}_j)$, and $H_{t,j}(\mathbf{r}_j)$ along all the conductor boundaries are the unknowns of the eigenmode problem and will in the sequel be referred to as *the sources*.

The integral equation used in this chapter is inspired by the success of the integral equation used for the wire transmission lines. We will make a distinction between the so-called internal regions inside the conductors and the so-called external regions outside the conductors. The fields in both the external and internal regions are expressed as a function of the sources at the conductor boundaries. We will again express the total field in terms of incoming and scattered fields. The fields in the external regions are the sum of incoming fields and scattered fields. The incoming fields are defined to be the fields generated by the sources if the layer, in which these sources are located, fills up the whole space. These incoming fields however will reflect at the layer interfaces and penetrate into the other layers. In this way the scattered fields are generated. The incoming fields exist only in the layer of the sources while the scattered fields exist in all layers. The internal and external regions are connected by imposing appropriate boundary conditions between the field expressions in both regions. This results in the final integral equation.

9.3.2 Internal regions

Consider the internal region of conductor j shown in Fig. 9.3. From Appendix A we know that the total longitudinal fields inside the conductor can be expressed as contour integrals along the boundary c_j:

$$E_{x,j}(\mathbf{r}) = E_{x,j}^{\text{in}}(\mathbf{r}) = \oint_{c_j} \left[E_{x,j}(\mathbf{r}'_j) \frac{\partial G_j(\mathbf{r}|\mathbf{r}'_j)}{\partial n'_j} - G_j(\mathbf{r}|\mathbf{r}'_j) \frac{\partial E_{x,j}(\mathbf{r}'_j)}{\partial n'_j} \right] dc'$$

$$H_{x,j}(\mathbf{r}) = H_{x,j}^{\text{in}}(\mathbf{r}) = \oint_{c_j} \left[H_{x,j}(\mathbf{r}'_j) \frac{\partial G_j(\mathbf{r}|\mathbf{r}'_j)}{\partial n'_j} - G_j(\mathbf{r}|\mathbf{r}'_j) \frac{\partial H_{x,j}(\mathbf{r}'_j)}{\partial n'_j} \right] dc'. \quad (9.1)$$

9.3 CONSTRUCTION OF THE INTEGRAL EQUATION

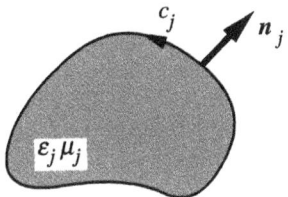

Fig. 9.3 Internal region of conductor j.

Because these total fields are analogous to the external incoming fields (9.2) discussed in the next section we explicitly introduced the superscript 'in' in (9.1). This will simplify the subsequent discussion. $G_j(\mathbf{r}|\mathbf{r}')$ is the two-dimensional Green's function of homogeneous space filled with the material of the conductor. From a knowledge of $E_{x,j}$, $H_{x,j}$ and their normal derivatives at the boundary c_j we can calculate $E_{x,j}$ and $H_{x,j}$ everywhere inside the conductor.

9.3.3 External regions

Now consider layer i in which some parts of conductor j are located (Fig. 9.4). First we concentrate on conductor part i_j. The incoming fields generated in layer i by this conductor part are given by

$$E_{x,i_j}^{\text{in}}(\mathbf{r}) = -\int_{c_{i_j}} \left[E_{x,i_j}(\mathbf{r}'_{i_j}) \frac{\partial G_i(\mathbf{r}|\mathbf{r}'_{i_j})}{\partial n'_{i_j}} - G_i(\mathbf{r}|\mathbf{r}'_{i_j}) \frac{\partial E_{x,i_j}(\mathbf{r}'_{i_j})}{\partial n'_{i_j}} \right] dc'$$

$$H_{x,i_j}^{\text{in}}(\mathbf{r}) = -\int_{c_{i_j}} \left[H_{x,i_j}(\mathbf{r}'_{i_j}) \frac{\partial G_i(\mathbf{r}|\mathbf{r}'_{i_j})}{\partial n'_{i_j}} - G_i(\mathbf{r}|\mathbf{r}'_{i_j}) \frac{\partial H_{x,i_j}(\mathbf{r}'_{i_j})}{\partial n'_{i_j}} \right] dc'. \quad (9.2)$$

$G_i(\mathbf{r}|\mathbf{r}')$ now represents the Green's function corresponding to layer i. Observe that the signs in (9.2) and (9.1) are different because the normal is now pointing inward relative to the external region. The incoming fields will

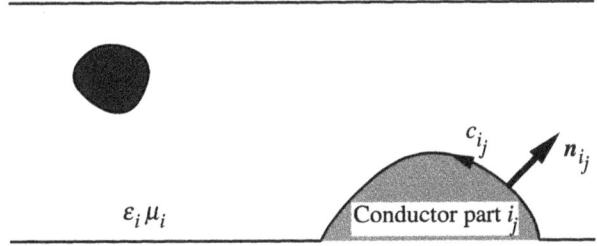

Fig. 9.4. External region around those parts of conductor j located in layer i.

generate scattered fields in all layers of the structure. Now we can write down the total fields in each layer i:

$$E_{x,i}(\mathbf{r}) = \sum_{j=1}^{C_i} E_{x,ij}^{\text{in}}(\mathbf{r}) + \sum_{j=1}^{C} E_{x,j}^{\text{sc}}(\mathbf{r})$$

$$H_{x,i}(\mathbf{r}) = \sum_{j=1}^{C_i} H_{x,ij}^{\text{in}}(\mathbf{r}) + \sum_{j=1}^{C} H_{x,j}^{\text{sc}}(\mathbf{r}).$$

(9.3)

$E_{x,j}^{\text{sc}}(\mathbf{r})$ and $H_{x,j}^{\text{sc}}(\mathbf{r})$ are the scattered fields generated in the layered structure by all the conductor parts of conductor j. If there are no conductor parts in layer i, the total fields in this layer will consist only of scattered fields due to the conductors in the other layers. Starting from the values of $E_{x,ij}$, $H_{x,ij}$, $\partial E_{x,ij}/\partial n$, and $\partial H_{x,ij}/\partial n$ at the conductor boundaries we can calculate the incoming fields, and from these the scattered fields, at every point in the external regions.

9.3.4 Boundary conditions and unknowns

In the previous two sections it was demonstrated that it is sufficient to know the longitudinal field components and their normal derivatives at the conductor boundaries to find the electromagnetic fields everywhere. This suggests the use of $E_x(\mathbf{r}_j)$, $H_x(\mathbf{r}_j)$, $\partial E_x(\mathbf{r}_j)/\partial n$, and $\partial H_x(\mathbf{r}_j)/\partial n$ as basic unknowns in the eigenmode problem. However, $\partial E_x(\mathbf{r}_j)/\partial n$ and $\partial H_x(\mathbf{r}_j)/\partial n$ are not suitable because they are not continuous and take different values at the inside and outside of the conductor boundaries.

To solve this problem we examine the continuity relations at the conductor boundaries. A first set of boundary conditions comes from the continuity of the longitudinal field components E_x and H_x. Secondly we have the continuity of the tangential field components E_t and H_t in the transverse (y, z) plane. It is easy to verify that other continuity relations between the normal field components are automatically fulfilled if E_x, H_x, E_t, and H_t are continuous.

In the local tangent coordinate system for curve c of Fig. 9.5 we can express E_t and H_t as a function of derivatives of E_x and H_x (cf. Appendix A):

Fig. 9.5. Local coordinate system for curve c.

9.3 CONSTRUCTION OF THE INTEGRAL EQUATION

$$E_t = -\frac{j\beta}{\gamma^2}\frac{\partial E_x}{\partial t} + \frac{j\omega\mu}{\gamma^2}\frac{\partial H_x}{\partial n}$$

$$H_t = -\frac{j\beta}{\gamma^2}\frac{\partial H_x}{\partial t} - \frac{j\omega\varepsilon}{\gamma^2}\frac{\partial E_x}{\partial n}.$$

(9.4)

By rearranging equation (9.4) we obtain expressions for the normal derivatives of E_x and H_x as a function of continuous quantities. Note, in this respect, that the continuity of E_x and H_x at the boundaries implies the continuity of their tangential derivatives. Consequently we will use E_t, H_t, $\partial E_x/\partial t$, and $\partial H_x/\partial t$ as unknowns instead of $\partial E_x/\partial n$ and $\partial H_x/\partial n$. If the functions E_x and H_x are known at the boundaries then $\partial E_x/\partial t$ and $\partial H_x/\partial t$ are also known by derivation along the boundary. Hence no new unknowns are introduced and the final set of unknown functions at the boundaries is E_x, H_x, E_t, and H_t.

Replacing the normal derivatives of the fields in (9.1), using (9.4), results in

$$E_{x,j}^{in}(\mathbf{r}) = \oint_{c_j}\left[E_{x,j}\frac{\partial G_j}{\partial n_j'} - G_j\left(\frac{j\gamma_j^2}{\omega\varepsilon_j}H_{t,j} - \frac{\beta}{\omega\varepsilon_j}\frac{\partial H_{x,j}}{\partial t_j'}\right)\right]dc'$$

$$H_{x,j}^{in}(\mathbf{r}) = \oint_{c_j}\left[H_{x,j}\frac{\partial G_j}{\partial n_j'} - G_j\left(-\frac{j\gamma_j^2}{\omega\mu_j}E_{t,j} + \frac{\beta}{\omega\mu_j}\frac{\partial E_{x,j}}{\partial t_j'}\right)\right]dc'$$

(9.5)

and an analogous replacement in equation (9.2) yields

$$E_{x,ij}^{in}(\mathbf{r}) = -\int_{c_{ij}}\left[E_{x,ij}\frac{\partial G_i}{\partial n_{ij}'} - G_i\left(\frac{j\gamma_i^2}{\omega\varepsilon_i}H_{t,ij} - \frac{\beta}{\omega\varepsilon_i}\frac{\partial H_{x,ij}}{\partial t_{ij}'}\right)\right]dc'$$

$$H_{x,ij}^{in}(\mathbf{r}) = -\int_{c_{ij}}\left[H_{x,ij}\frac{\partial G_i}{\partial n_{ij}'} - G_i\left(-\frac{j\gamma_i^2}{\omega\mu_i}E_{t,ij} + \frac{\beta}{\omega\mu_i}\frac{\partial E_{x,ij}}{\partial t_{ij}'}\right)\right]dc'.$$

(9.6)

9.3.5 Integral equation

The integral equations are constructed by connecting the internal and external regions. This means that we impose the continuity relations between the total fields at the outside and the total fields at the inside of the conductor boundaries. As explained in the previous section, it is sufficient to impose the continuity of the longitudinal components E_x and H_x and the tangential transversal components E_t and H_t:

$$\lim_{\mathbf{r} \gtrless \mathbf{r}_{i_o}} \left[\sum_{j=1}^{C_i} E^{\text{in}}_{x,i_j}(\mathbf{r}) + \sum_{j=1}^{C} E^{\text{sc}}_{x,j}(\mathbf{r}) \right] = \lim_{\mathbf{r} \gtrless \mathbf{r}_{i_o}} E^{\text{in}}_{x,o}(\mathbf{r})$$

$$\lim_{\mathbf{r} \gtrless \mathbf{r}_{i_o}} \left[\sum_{j=1}^{C_i} H^{\text{in}}_{x,i_j}(\mathbf{r}) + \sum_{j=1}^{C} H^{\text{sc}}_{x,j}(\mathbf{r}) \right] = \lim_{\mathbf{r} \gtrless \mathbf{r}_{i_o}} H^{\text{in}}_{x,o}(\mathbf{r})$$

$$\lim_{\mathbf{r} \gtrless \mathbf{r}_{i_o}} \left[\sum_{j=1}^{C_i} E^{\text{in}}_{t,i_j}(\mathbf{r}) + \sum_{j=1}^{C} E^{\text{sc}}_{t,j}(\mathbf{r}) \right] = \lim_{\mathbf{r} \gtrless \mathbf{r}_{i_o}} E^{\text{in}}_{t,o}(\mathbf{r})$$
(9.7)

$$\lim_{\mathbf{r} \gtrless \mathbf{r}_{i_o}} \left[\sum_{j=1}^{C_i} H^{\text{in}}_{t,i_j}(\mathbf{r}) + \sum_{j=1}^{C} H^{\text{sc}}_{t,j}(\mathbf{r}) \right] = \lim_{\mathbf{r} \gtrless \mathbf{r}_{i_o}} H^{\text{in}}_{t,o}(\mathbf{r})$$

$$i_o = 1, \ldots, C_i \qquad i = 1, \ldots, L.$$

$\mathbf{r}_{i_o}$ is a position vector located on the oth conductor part in layer i. The subscript 'o' (from the word observation) in the incoming fields on the right-hand side refers to the conductor to which conductor part i_o belongs. The limiting operation at the left-hand side is performed from the external region to the conductor boundary as indicated by the notation '$\mathbf{r} \gtrless \mathbf{r}_{i_o}$' and the operation at the right-hand side is performed from the internal region as indicated by the notation '$\mathbf{r} \gtrless \mathbf{r}_{i_o}$'. In the last two equations of (9.7) the tangential transversal fields (index t) are determined from their longitudinal counterparts (index x) by applying (9.4). The integral equations (9.7) consist of four continuity relations between $E_x(\mathbf{r})$ and $H_x(\mathbf{r})$ in four unknown functions $E_x(\mathbf{r}_j)$, $H_x(\mathbf{r}_j)$, $E_t(\mathbf{r}_j)$, and $H_t(\mathbf{r}_j)$. These unknowns are defined over the total boundary of all conductors j, $j = 1, \ldots, C$. Equations (9.7) are written down on a per-layer basis. For that reason the layer index i counts from 1 to L, where L is the number of layers in the stratified medium. In the following we refer to (9.7) as the *integral equation*.

9.3.6 *Perfectly conducting conductors*

The case of perfectly conducting conductors allows some simplifications and makes the connection with the previous chapter clear.

Inside perfect conductors the electromagnetic fields vanish, so we do not have to consider the internal regions. The electric field is perpendicular to the boundary of a perfect conductor. This means that $E_x(\mathbf{r}_j) = E_t(\mathbf{r}_j) = 0$, or that only two unknown functions $H_x(\mathbf{r}_j)$ and $H_t(\mathbf{r}_j)$ remain at the perfectly conducting conductor boundaries. At these boundaries we can impose only two conditions: $E_x(\mathbf{r} \gtrless \mathbf{r}_j) = E_t(\mathbf{r} \gtrless \mathbf{r}_j) = 0$. Hence equation (9.7) reduces to

9.4 SOLUTION TECHNIQUE FOR THE INTEGRAL EQUATION

a set of two equations with two unknown functions:

$$\lim_{\mathbf{r} \gtrless \mathbf{r}_{i_o}} \left[\sum_{j=1}^{C_i} E_{x,i,j}^{in}(\mathbf{r}) + \sum_{j=1}^{C} E_{x,j}^{sc}(\mathbf{r}) \right] = 0 \tag{9.8}$$

$$\lim_{\mathbf{r} \gtrless \mathbf{r}_{i_o}} \left[\sum_{j=1}^{C_i} E_{t,i,j}^{in}(\mathbf{r}) + \sum_{j=1}^{C} E_{t,j}^{sc}(\mathbf{r}) \right] = 0$$

$$i_o = 1, \ldots, C_i \qquad i = 1, \ldots, L$$

and equation (9.6) reduces to

$$E_{x,ij}^{in} = \int_{c_{ij}} G_i \left(\frac{j\gamma_i^2}{\omega \varepsilon_i} H_{t,ij} - \frac{\beta}{\omega \varepsilon_i} \frac{\partial H_{x,ij}}{\partial t'_{ij}} \right) dc'$$

$$H_{x,ij}^{in} = -\int_{c_{ij}} H_{x,ij} \frac{\partial G_i}{\partial n'_{ij}} dc'. \tag{9.9}$$

The integral equation (9.8) and the final expression (9.9) are slightly different from the ones found in the previous chapter. In the previous chapter we used the longitudinal magnetic field or the tangential surface current and the normal derivative of the longitudinal electric field as unknowns. In (9.9) this normal derivative was replaced, using equation (9.4), as explained in Section 9.3.4. In the previous chapter this replacement was not necessary because there was no connection with the internal regions. There we imposed conditions for the longitudinal electric and magnetic fields at the conductors. In (9.8) the condition for the magnetic field is replaced by an equivalent condition for the transversal tangential electric field.

9.4 Solution technique for the integral equation

The integral equation (9.7) is solved by Galerkin's method. This technique transforms the original coupled set of equations (9.7) into a linear system of equations in the unknown amplitudes of the basis functions. This system forms an eigenvalue problem in which the propagation constants β of the modes are the eigenvalues.

9.4.1 *Basis functions and test functions*

The basis and test functions in the moment method of Galerkin will be chosen in a natural way. We first concentrate on the basis functions. At a perfect conductor, H_x is equal to the transversal surface current on the conductor boundary. In order to avoid non-physical charge concentrations we need a continuous representation for H_x. The simplest continuous representation is a piecewise linear representation. H_t corresponds to the longitudinal surface current, for which no continuous representation is needed. We will use a piecewise constant representation for H_t. For a conductor with finite conductivity we adopt the same choices. If we look at the first equation of (9.5) we see a sum of a term proportional to H_t and a term proportional to $\partial H_x/\partial t$. When H_x is piecewise linear, $\partial H_x/\partial t$ will be piecewise constant; so it would indeed not make much sense to use a higher-order representation for H_t. In [6] and [7] the same choice of basis functions is used for the representation of the surface current. By analogy we use a piecewise linear representation for E_x and a piecewise constant representation for E_t.

From (9.4) we see that the behaviour of E_t and H_t is one order more singular than the behaviour of E_x and H_x due to the derivatives. This suggests that we need higher-order test functions for the E_t and H_t than for the E_x and H_x boundary conditions. In order to enjoy the benefits of Galerkin's procedure we use piecewise constant test functions for E_x and H_x and piecewise linear functions for E_t and H_t.

This choice of basis and test functions results in a very elegant and consistent development of the theory. We will come back to this point later.

We use the standard piecewise constant and piecewise linear representation technique. The boundary of the conductor (Fig. 9.6) is divided into a number of segments with not necessarily equal length. For the piecewise constant representation we use pulse functions $p(\tau)$ where τ is the arc length (Fig. 9.6(b)):

$$p(\tau) = \frac{1}{\delta} \qquad 0 \leqslant \tau \leqslant \delta. \tag{9.10}$$

Overlapping triangular functions $t(\tau)$ are used for the piecewise linear representation (Fig. 9.6(c)):

$$t(\tau) = \begin{cases} 1 + \dfrac{\tau}{\delta_1} & -\delta_1 \leqslant \tau \leqslant 0 \\ 1 - \dfrac{\tau}{\delta_2} & 0 \leqslant \tau \leqslant \delta_2. \end{cases} \tag{9.11}$$

The total number of segments on all conductor boundaries is equal to N. $p_k(\tau)$, $\tau \in [0, \delta_k]$ ($k = 1, \ldots, N$), denotes the kth pulse function and $t_k(\tau)$, $\tau \in [-\delta_{1,k}, \delta_{2,k}]$ ($k = 1, \ldots, N$), denotes the kth triangular function.

The segment on a conductor corresponding to a specific basis function

9.4 SOLUTION TECHNIQUE FOR THE INTEGRAL EQUATION

will be called the excitation segment. A segment corresponding to a specific test function will be called the observation segment. So each segment is sometimes seen as an excitation segment and sometimes as an observation segment depending on the context.

9.4.2 Discrete eigenvalue problem

If the total number of segments on all conductors is equal to N, our integral equation is discretized in a set of $4N$ homogeneous linear equations, each corresponding to a test function, with $4N$ unknown amplitudes of the basis functions:

$$\begin{bmatrix}
[E_xE_x]_{11} & \cdots & [E_xE_x]_{1N} & [E_xH_t]_{11} & \cdots & [E_xH_t]_{1N} \\
\vdots & & \vdots & \vdots & & \vdots \\
[E_xE_x]_{N1} & \cdots & [E_xE_x]_{NN} & [E_xH_t]_{N1} & \cdots & [E_xH_t]_{NN} \\
[H_xE_x]_{11} & \cdots & [H_xE_x]_{1N} & [H_xH_t]_{11} & \cdots & [H_xH_t]_{1N} \\
\vdots & & \vdots & \vdots & & \vdots \\
[H_xE_x]_{N1} & \cdots & [H_xE_x]_{NN} & [H_xH_t]_{N1} & \cdots & [H_xH_t]_{NN} \\
[E_tE_x]_{11} & \cdots & [E_tE_x]_{1N} & [E_tH_t]_{11} & \cdots & [E_tH_t]_{1N} \\
\vdots & & \vdots & \vdots & & \vdots \\
[E_tE_x]_{N1} & \cdots & [E_tE_x]_{NN} & [E_tH_t]_{N1} & \cdots & [E_tH_t]_{NN} \\
[H_tE_x]_{11} & \cdots & [H_tE_x]_{1N} & [H_tH_t]_{11} & \cdots & [H_tH_t]_{1N} \\
\vdots & & \vdots & \vdots & & \vdots \\
[H_tE_x]_{N1} & \cdots & [H_tE_x]_{NN} & [H_tH_t]_{N1} & \cdots & [H_tH_t]_{NN}
\end{bmatrix}
\begin{bmatrix} E_{x,1} \\ \vdots \\ E_{x,N} \\ H_{x,1} \\ \vdots \\ H_{x,N} \\ E_{t,1} \\ \vdots \\ E_{t,N} \\ H_{t,1} \\ \vdots \\ H_{t,N} \end{bmatrix}
= \begin{bmatrix} 0 \\ \vdots \\ 0 \\ 0 \\ \vdots \\ 0 \\ 0 \\ \vdots \\ 0 \\ 0 \\ \vdots \\ 0 \end{bmatrix}.$$

(9.12)

$E_{x,k}$, $H_{x,k}$, $E_{t,k}$, and $H_{t,k}$ ($k = 1, \ldots, N$) are the unknown amplitudes of the basis functions in the representation of E_x, H_x, E_t, and H_t. $[XY]_{kl}$ ($k, l = 1, \ldots, N$; $X, Y = E_x, H_x, E_t, H_t$) is the X boundary condition tested at the kth observation segment and generated by the basis function of the field component Y at the lth excitation segment. So $[XY]_{kl}$ contains an integration over the lth excitation segment and over the kth observation segment. Comparison of (9.7) and (9.12) shows that $[XY]_{kl}$ is formed by the difference of an internal and external part and that this external part consists of an incoming and scattered contribution:

$$[XY]_{kl} = [XY]_{kl}^{\text{external}} - [XY]_{kl}^{\text{internal}}$$
$$= [XY]_{kl}^{\text{external, in}} + [XY]_{kl}^{\text{external, sc}} - [XY]_{kl}^{\text{internal, in}}.$$

(9.13)

214   9 POLYGONAL TRANSMISSION LINE STRUCTURES

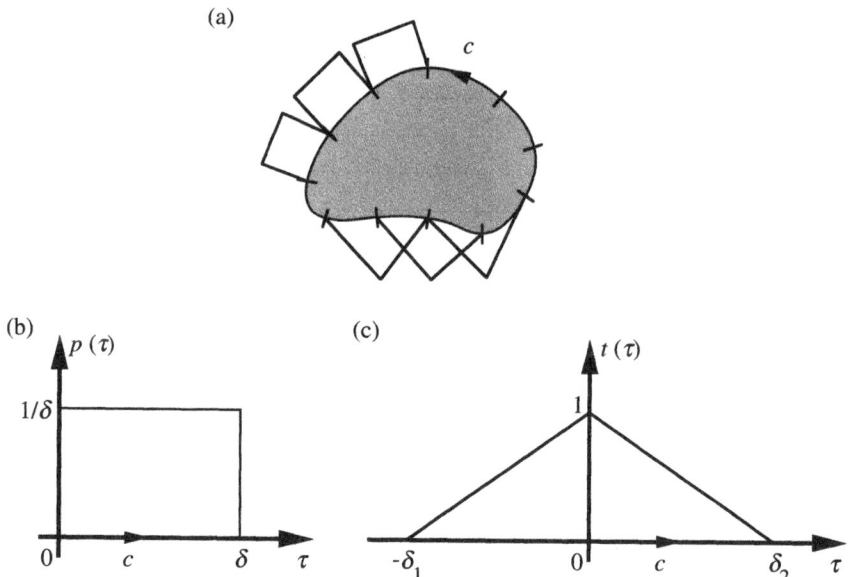

Fig. 9.6. Conductor boundary divided into a number of segments: (a) conductor; (b) pulse function; (c) triangular function.

Each coefficient $[XY]_{kl}$ in (9.12) is a non-linear function of the propagation constant β. Equation (9.12) will have only non-trivial solutions if the determinant of the system matrix becomes zero. This determinant is a function of β and will become zero for some discrete values of β. These values correspond to the propagation constants of the modes and the corresponding solutions or eigenvectors of (9.12) are the tangential field components at the conductor boundaries of the modes.

9.4.3 The system matrix

9.4.3.1 Space domain
In this section we will give explicit expressions for the excitation and observation integration appearing in each coefficient of the system matrix. First the observation integration over the test function is examined followed by the excitation integration.

Each coefficient of type $[E_x Y]_{kl}$, with $Y = E_x$, E_t, H_x, or H_t, contains an observation integration of E_x over the kth segment. This integration takes the form

$$\int_0^{\delta_k} p_k(v) E_x(\mathbf{r}(v)) \, dv \tag{9.14}$$

where $E_x(\mathbf{r})$ is the longitudinal electric field generated by the lth basis

9.4 SOLUTION TECHNIQUE FOR THE INTEGRAL EQUATION

function of field component Y. $E_x(\mathbf{r})$ can be either an incoming or a scattered field. We use the integration variable v to indicate an integration over an observation interval. Analogously, $[H_x Y]_{kl}$ contains an integration of the form

$$\int_0^{\delta_k} p_k(v) H_x(\mathbf{r}(v)) \, dv. \tag{9.15}$$

In $[E_t Y]_{kl}$, on the other hand, we have the following integration:

$$\int_{-\delta_{k,1}}^{\delta_{k,2}} t_k(v) E_t(\mathbf{r}(v)) \, dv. \tag{9.16}$$

The transversal tangential field $E_t(\mathbf{r})$ has to be expressed in terms of E_x and H_x using (9.4):

$$[E_t Y]_{kl} = \int_{-\delta_{k,1}}^{\delta_{k,2}} t_k(v) \left[-\frac{j\beta}{\gamma^2} \frac{\partial E_x(\mathbf{r}(v))}{\partial t} + \frac{j\omega\mu}{\gamma^2} \frac{\partial H_x(\mathbf{r}(v))}{\partial n} \right] dv \tag{9.17}$$

and finally $[H_t Y]_{kl}$ can be written as

$$[H_t Y]_{kl} = \int_{-\delta_{k,1}}^{\delta_{k,2}} t_k(v) \left[-\frac{j\beta}{\gamma^2} \frac{\partial H_x(\mathbf{r}(v))}{\partial t} - \frac{j\omega\varepsilon}{\gamma^2} \frac{\partial E_x(\mathbf{r}(v))}{\partial n} \right] dv \tag{9.18}$$

Each coefficient of type $[XE_x]_{kl}$, with $X = E_x, E_t, H_x$, or H_t, contains an excitation integration over the lth basis function of the representation of E_x. This integration generates the longitudinal incoming fields needed directly or indirectly to calculate an incoming or scattered field contribution. The integrations associated with $[XE_x]_{kl}$ are

$$E_x^{\text{in}}(\mathbf{r}) = \int_{-\delta_{l,1}}^{\delta_{l,2}} t_l(\tau) \frac{\partial G(\mathbf{r}|\mathbf{r}'(\tau))}{\partial n'} \, d\tau \tag{9.19}$$

and

$$H_x^{\text{in}}(\mathbf{r}) = -\frac{\beta}{\omega\mu} \int_{-\delta_{l,1}}^{\delta_{l,2}} \frac{dt_l(\tau)}{d\tau} G(\mathbf{r}|\mathbf{r}'(\tau)) \, d\tau \tag{9.20}$$

as can be verified directly from (9.5). Here we used the integration variable τ to indicate an integration over an excitation segment. Analogously for $[XH_x]_{kl}$:

$$E_x^{\text{in}}(\mathbf{r}) = \frac{\beta}{\omega\varepsilon} \int_{-\delta_{l,1}}^{\delta_{l,2}} \frac{dt_l(\tau)}{d\tau} G(\mathbf{r}|\mathbf{r}'(\tau)) \, d\tau \tag{9.21}$$

and

$$H_x^{in}(\mathbf{r}) = \int_{-\delta_{l,1}}^{\delta_{l,2}} t_l(\tau) \frac{\partial G(\mathbf{r}|\mathbf{r}'(\tau))}{\partial n'} \, d\tau. \quad (9.22)$$

From (9.5) one sees that $[XE_t]_{kl}$ has only a magnetic contribution:

$$H_x^{in}(\mathbf{r}) = \frac{j\gamma^2}{\omega\mu} \int_0^{\delta_l} p_l(\tau) G(\mathbf{r}|\mathbf{r}'(\tau)) \, d\tau \quad (9.23)$$

and finally for $[XH_t]_{kl}$ which has only an electric contribution:

$$E_x^{in}(\mathbf{r}) = -\frac{j\gamma^2}{\omega\varepsilon} \int_0^{\delta_l} p_l(\tau) G(\mathbf{r}|\mathbf{r}'(\tau)) \, d\tau. \quad (9.24)$$

In equations (9.19)–(9.24) we started from equation (9.5) for incoming fields in the internal regions. However for the external regions, that is, equation (9.6), an extra minus sign has to be introduced in (9.19)–(9.24).

9.4.3.2 *Spectral domain* In (9.19)–(9.24) the basis functions t_l (and its derivative) and p_l are multiplied by the Green's function $G(\mathbf{r}|\mathbf{r}') = j/4 H_0^{(2)}(\gamma|\mathbf{r} - \mathbf{r}'|)$ (or its derivative), leading to complicated integrations. This problem can be solved if we spatially Fourier transform the expressions (9.19)–(9.24) and calculate the spectral longitudinal incoming fields. Interchanging the Fourier transform with the excitation integration and possible normal derivatives (appearing in (9.19) and (9.22)) leads us to the Fourier transform of the Green's function. This Fourier transform is given by [13]:

$$\frac{1}{2\pi} \int_{-\infty}^{+\infty} G(\mathbf{r}|\mathbf{r}') e^{jk_y y} \, dy = -\frac{1}{4\pi} \frac{e^{jk_y y'} e^{-\Gamma|z-z'|}}{\Gamma}. \quad (9.25)$$

The introduction of this spectral Green's function allows us to perform the excitation integration analytically if the excitation segment is a straight line as shown in Section 9.5.2. In analogy with the previous chapter we also need the spectral incoming fields in the calculation of the scattered fields.

The space domain fields appearing in the test integrations (9.14), (9.15), (9.17), and (9.18) are found by inverse Fourier transforming the incoming or scattered spectral fields. We will however first bring the inverse Fourier transform outside the observation integration by interchanging both integrations. This makes it possible to perform the observation integration (that is, the integration over v) analytically if the observation segment is a straight line. Applying the sketched procedure, the equations (9.14), (9.15), (9.17), and (9.18) now take the following canonical form:

$$[XY]_{kl} = \int_{-\infty}^{+\infty} \int u_k(v) X(k_y, z(v)) e^{-jk_y y(v)} \, dv \, dk_y \quad (9.26)$$

9.4 SOLUTION TECHNIQUE FOR THE INTEGRAL EQUATION

where $u(v)$ is either $p(v)$ or $t(v)$. The observation integration over v can be performed analytically as shown in Section 9.5.3.

In conclusion it can be stated that the spectral transformation has two benefits. Firstly the scattered fields can be calculated analytically in the Fourier domain (as in Chapter 8) and secondly the excitation and observation integrations can be performed analytically for polygonal conductors. Only the final inverse Fourier transform remains to be integrated numerically with the techniques described in Chapter 6. Note that we use this spectral technique for internal regions with finite dimensions, and not only for the surrounding stratified medium.

9.4.4 Incoming coefficients

Even after applying the spectral domain technique the excitation and observation integrations can become very complicated. This is due to the absolute value of $|z - z'|$ appearing in the spectral Green's function (9.25). Consider a typical integration over an excitation segment. If we need the spectral incoming fields at a z-coordinate above or below the excitation segment there is no problem because $(z - z')$ has a fixed sign. However, if we need the spectral incoming field at a z-coordinate at the height of the excitation segment we have to divide the integration interval into two parts: one with $|z - z'| = (z - z')$ and one with $|z - z'| = -(z - z')$. At the observation segment we integrate over z. This means that if the observation and excitation segments have overlapping z-coordinates, as shown in Fig. 9.7(a), we have to divide both integrations into several parts. In region I $(z - z')$ is positive, in region III $(z - z')$ is negative, and in region II $(z - z')$ can have both signs and the excitation and observation integrations are coupled. However, this problem can be solved in an elegant way.

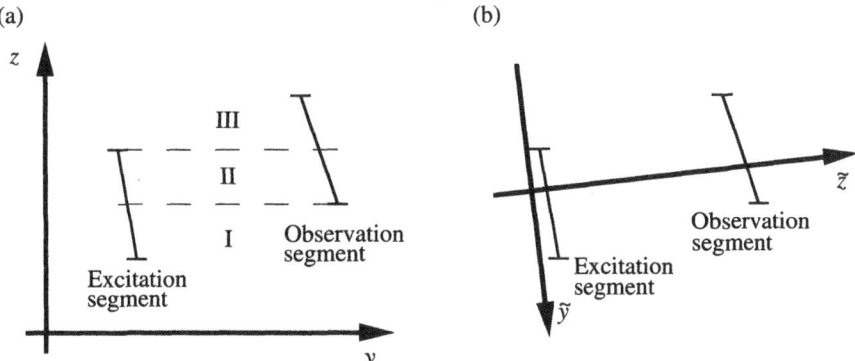

Fig. 9.7. Observation and excitation segment: (a) with overlapping z-coordinates in the (y, z) coordinate system; (b) without overlapping $\tilde{z}$-coordinates in the $(\tilde{y}, \tilde{z})$ coordinate system.

The incoming fields are fields generated in a homogeneous space of infinite extent, so there is no preferential spectral transformation direction imposed by any layered medium. If we use the new coordinate system $(\tilde{y}, \tilde{z})$ of Fig. 9.7(b) and if we Fourier transform along the new $\tilde{y}$-direction, this partitioning in zones is no longer necessary because $(\tilde{z} - \tilde{z}')$ is always positive. In some special configurations it is impossible to construct an appropriate $(\tilde{y}, \tilde{z})$ coordinate system. However, this problem can always be solved by dividing the conductor boundaries into smaller segments. One problem case remains when the excitation and observation segments coincide, that is, the so-called selfpatch contribution. This problem can be solved in an elegant way only for polygonal waveguides, in which case the $\tilde{y}$-axis is taken along the straight segment and $(\tilde{z} - \tilde{z}')$ is equal to zero. In the further calculation we will always use the standard (y, z) coordinate system notation, but for the incoming coefficients this system is assumed to be the $(\tilde{y}, \tilde{z})$ coordinate system. The reader could object that there can be a problem when combining a triangular test function $t(v)$ with a triangular basis function $t(\tau)$ at a corner of the conductor. In this case we use two different $(\tilde{y}, \tilde{z})$ coordinate systems, one for the part $\tau > 0$ and one for the part $\tau < 0$.

9.4.5 Scattered coefficients

For the scattered coefficients we proceed in an analogous way as in the previous chapter. We will show that the spectral longitudinal incoming fields generated by each basis function can be cast into the following general form (see Section 9.2.5):

$$E_x^{\text{in}}(k_y, z) = \tilde{e}^+(k_y) \, e^{-\Gamma_e z}$$
$$H_x^{\text{in}}(k_y, z) = \tilde{h}^+(k_y) \, e^{-\Gamma_e z} \quad (9.27)$$

for z values in the excitation layer above the excitation segment. On the other hand for z values under the excitation segment this becomes

$$E_x^{\text{in}}(k_y, z) = \tilde{e}^-(k_y) \, e^{\Gamma_e z}$$
$$H_x^{\text{in}}(k_y, z) = \tilde{h}^-(k_y) \, e^{\Gamma_e z} \quad (9.28)$$

Equations (9.27) (respectively (9.28)) are equal to equations (8.36) (respectively (8.37)) with

$$e^+(k_y) = \tilde{e}^+(k_y) \, e^{\Gamma_e z_{e,e+1}}$$
$$h^+(k_y) = \tilde{h}^+(k_y) \, e^{\Gamma_e z_{e,e+1}} \quad (9.29)$$

9.5 EXCITATION AND OBSERVATION INTEGRATIONS

(respectively

$$e^-(k_y) = \tilde{e}^-(k_y) \, e^{-\Gamma_e z_{e-1,e}}$$
$$h^-(k_y) = \tilde{h}^-(k_y) \, e^{-\Gamma_e z_{e-1,e}}). \tag{9.30}$$

This means that the scattered fields can be calculated using exactly the same technique as in Section 8.6. We will not repeat this calculation here and assume immediately that we have the spectral longitudinal scattered fields in each layer i in the following general form:

$$E_x^{sc}(k_y, z) = A_e^{sc} \, e^{-\Gamma_i(z - z_{i-1,i})} + B_e^{sc} \, e^{-\Gamma_i(z_{i,i+1} - z)}$$
$$H_x^{sc}(k_y, z) = A_h^{sc} \, e^{-\Gamma_i(z - z_{i-1,i})} + B_h^{sc} \, e^{-\Gamma_i(z_{i,i+1} - z)}. \tag{9.31}$$

This is analogous to equations (8.102) and (8.104) but with a change of reference z-coordinate from the centre of the wire to the top and bottom of layer i.

9.5 Excitation and observation integrations

9.5.1 *Polygonal conductors*

In the further development of the theory it is necessary to assume conductors with polygonal cross-section. As mentioned above, this will allow us to perform the basis and test function integrations analytically. Figure 9.8(a) represents a polygonal conductor with its boundary divided into segments. Corners of the boundary are always at the edges of segments. If a polygonal side crosses the boundary between two dielectric layers a segment edge is always introduced at the layer interface.

Figure 9.8(b) shows some notations used to handle pulse basis and test functions. The begin and end point of the segment respectively have coordinates (y_0, z_0) and (y_1, z_1) and the segment makes an angle α with the y-axis. The triangular function case is shown in Fig. 9.8(c). The point at the top of the triangle has coordinates (y_0, z_0) and the left and right end points have coordinates (y_1, z_1) and (y_2, z_2). Two angles α_1 and α_2 are introduced because a triangular segment can run across a corner of a conductor. If no corner is present $\alpha_1 = \alpha_2$.

9.5.2 *Excitation integration*

In this section we perform the excitation integration, in other words we determine the spectral longitudinal incoming fields generated by each basis function. This means that the Fourier transforms that appear in (9.19)–(9.24) are calculated by the procedure described in Section 9.4.3.

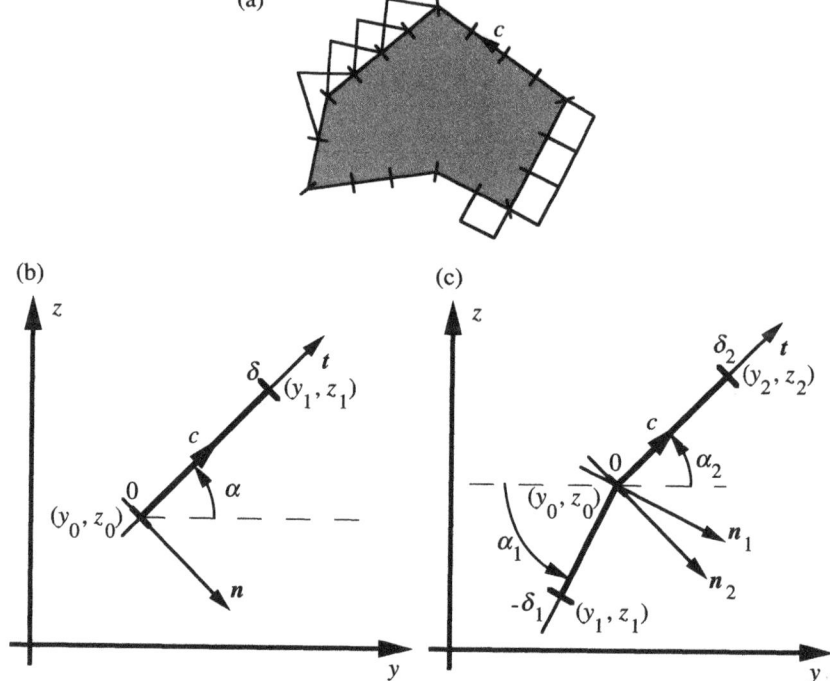

Fig. 9.8. Polygonal conductor: (a) polygonal boundary divided into segments; (b) a pulse basis or test function; (c) a triangular basis or test function.

In (9.19)–(9.24) and in this section we use some material-dependent constants such as ε, μ, γ, and Γ. For external incoming fields these constants correspond to the material of the layer in which the fields are calculated. For internal fields the material of the conductor is assumed. We will use the conditions $z > z'$ (respectively $z < z'$), by which we mean that z must be taken above (respectively below) all the z'-coordinates of the excitation segment.

We start with the contribution of a basis function of H_t to the longitudinal incoming electric field $E_x^{\text{in}}(k_y, z)$ as appearing in (9.24). To simplify the notation the subscript l, referring to a specific basis function, is omitted. With the spectral Green's function (9.25) and the pulse function (9.10) one finds, when $z > z'$:

$$E_x^{\text{in}}(k_y, z) = \frac{j\gamma^2 e^{-\Gamma z}}{4\pi \Gamma \varepsilon \omega \delta} \int_0^\delta e^{jk_y y'} e^{\Gamma z'} \, d\tau. \tag{9.32}$$

y' and z' are linear functions in τ:

$$\begin{aligned} y' &= y_0 + \tau \cos \alpha \\ z' &= z_0 + \tau \sin \alpha. \end{aligned} \tag{9.33}$$

9.5 EXCITATION AND OBSERVATION INTEGRATIONS

Inserting (9.33) in (9.32) and integrating yields

$$E_x^{in}(k_y, z) = \frac{j\gamma^2 \, e^{-\Gamma z + m_0}(e^{\delta t} - 1)}{4\pi\Gamma\varepsilon\omega\delta t} \tag{9.34}$$

with m_0 and t given by

$$m_0 = jk_y y_0 + \Gamma z_0$$
$$t = jk_y \cos\alpha + \Gamma \sin\alpha. \tag{9.35}$$

When $z < z'$ the reader can verify that (9.34) is replaced by

$$E_x^{in}(k_y, z) = \frac{j\gamma^2 \, e^{\Gamma z + n_0}(e^{\delta u} - 1)}{4\pi\Gamma\varepsilon\omega\delta u} \tag{9.36}$$

where

$$n_0 = jk_y y_0 - \Gamma z_0$$
$$u = jk_y \cos\alpha - \Gamma \sin\alpha. \tag{9.37}$$

Now we calculate the $E_x^{in}(k_y, z)$ field generated by an H_x triangular basis function (9.11) as appearing in (9.21). Differentiating (9.11) with respect to τ results in

$$\frac{dt(\tau)}{d\tau} = \begin{cases} \dfrac{1}{\delta_1} & -\delta_1 \leqslant \tau \leqslant 0 \\ -\dfrac{1}{\delta_2} & 0 \leqslant \tau \leqslant \delta_2. \end{cases} \tag{9.38}$$

After integration we get, for $z > z'$:

$$E_x^{in}(k_y, z) = \frac{\beta \, e^{-\Gamma z + m_0}}{4\pi\Gamma\varepsilon\omega}\left[\frac{e^{-\delta_1 t_1} - 1}{\delta_1 t_1} + \frac{e^{\delta_2 t_2} - 1}{\delta_2 t_2}\right]. \tag{9.39}$$

t_1 (respectively t_2) correspond to t in (9.35) with α replaced by α_1 (respectively α_2). This means that in the case of a straight excitation segment, with $\alpha_1 = \alpha_2$, t_1 is equal to t_2. The derivation of the result for $z < z'$ is left to the reader.

Now we concentrate on the E_x basis function contribution (9.19) to $E_x^{in}(k_y, z)$. In this case we need the normal derivative of the spectral Green's function. For this we use the following relations between the (t, n) and (y, z) derivatives (cf. Appendix A):

$$\frac{\partial}{\partial n} = \sin\alpha \frac{\partial}{\partial y} - \cos\alpha \frac{\partial}{\partial z}$$
$$\frac{\partial}{\partial t} = \cos\alpha \frac{\partial}{\partial y} + \sin\alpha \frac{\partial}{\partial z}. \tag{9.40}$$

For $z > z'$ the Fourier transform of the normal derivative of G is equal to

$$\frac{1}{2\pi} \int_{-\infty}^{+\infty} \frac{\partial G(\mathbf{r}|\mathbf{r}')}{\partial n'} e^{jk_y y} \, dy = -\frac{w}{4\pi} \frac{e^{jk_y y'} e^{-\Gamma(z-z')}}{\Gamma} \quad (9.41)$$

where

$$w = jk_y \sin \alpha - \Gamma \cos \alpha. \quad (9.42)$$

$E_x^{\text{in}}(k_y, z)$ with (9.19) and (9.11) is given by

$$E_x^{\text{in}}(k_y, z) = -\frac{e^{-\Gamma z}}{4\pi\Gamma} \left[w_1 \int_{-\delta_1}^{0} \left(1 + \frac{\tau}{\delta_1}\right) e^{jk_y y'} e^{\Gamma z'} \, d\tau \right.$$

$$\left. + w_2 \int_{0}^{\delta_2} \left(1 - \frac{\tau}{\delta_2}\right) e^{jk_y y'} e^{\Gamma z'} \, d\tau \right]. \quad (9.43)$$

If we integrate by parts we finally find

$$E_x^{\text{in}}(k_y, z) = -\frac{e^{-\Gamma z + mo}}{4\pi\Gamma} \left[\frac{w_1(e^{-\delta_1 t_1} - 1)}{\delta_1 t_1^2} + \frac{w_2(e^{\delta_2 t_2} - 1)}{\delta_2 t_2^2} + \frac{w_1}{t_1} - \frac{w_2}{t_2} \right]. \quad (9.44)$$

Note that for a straight excitation segment, without a corner at $\tau = 0$, t_1 and w_1 are respectively equal to t_2 and w_2. This means that in this case the last two terms in (9.44) cancel out. The derivation of the $z < z'$ case is again left to the reader.

The calculation of the spectral longitudinal magnetic incoming field $H_x^{\text{in}}(k_y, z)$ is analogous to the calculation of $E_x^{\text{in}}(k_y, z)$. From (9.34), (9.36), (9.39), and (9.44) we see that $E_x^{\text{in}}(k_y, z)$, and by analogy $H_x^{\text{in}}(k_y, z)$, are of the general form (9.27) and (9.28) as put forward in Section 9.4.5.

Again a minus sign has to be introduced in (9.32), (9.34), (9.36), (9.39), (9.43), and (9.44) for incoming fields in the external regions. A last remark is that one only needs to consider the $z > z'$ case for the calculation of the spectral incoming coeffcients due to the special orientation of the $(\tilde{y}, \tilde{z})$ coordinate system.

9.5.3 Observation integration

In this section we will perform the observation integration. This can be an integration over incoming field or scattered field contributions. Subscripts referring to conductors or layers are again omitted. The longitudinal spectral incoming or scattered fields can be written as a sum of terms of the following general form:

$$E_x^-(k_y, z) = A_e(k_y) \, e^{-\Gamma z}$$
$$H_x^-(k_y, z) = A_h(k_y) \, e^{-\Gamma z} \quad (9.45)$$

9.5 EXCITATION AND OBSERVATION INTEGRATIONS

and

$$E_x^+(k_y, z) = B_e(k_y) \, e^{\Gamma z}$$
$$H_x^+(k_y, z) = B_h(k_y) \, e^{\Gamma z}. \tag{9.46}$$

For incoming fields this form is evident from the previous section or from (9.27). Note that due to the special choice of the $(\tilde{y}, \tilde{z})$ coordinate system only incoming fields of the form (9.45) need to be considered. The scattered fields in (9.31) each contain two terms of the form (9.45) and (9.46).

Due to the interchange of the inverse Fourier transform and the observation integration described in Section 9.4.3, we have to multiply (9.45) and (9.46) by the y dependency (cf. equation (9.26)) from the inverse Fourier transformation before performing the observation integration, yielding

$$E_x^-(y, z, k_y) = A_e(k_y) \, e^{-\Gamma z} \, e^{-jk_y y}$$
$$H_x^-(y, z, k_y) = A_h(k_y) \, e^{-\Gamma z} \, e^{-jk_y y} \tag{9.47}$$

and

$$E_x^+(y, z, k_y) = B_e(k_y) \, e^{\Gamma z} \, e^{-jk_y y}$$
$$H_x^+(y, z, k_y) = B_h(k_y) \, e^{\Gamma z} \, e^{-jk_y y}. \tag{9.48}$$

Integrating (9.47) and (9.48) over k_y from $]-\infty, +\infty[$ would result in the spatial longitudinal fields. However we will first perform the observation integration. In the following only the up-propagating waves of equation (9.47) will be considered; the down-propagating waves in (9.48) are left to the reader.

First we consider the testing of the E_x field boundary condition; this is equation (9.14). This involves the testing of (9.47) with the pulse function (9.10):

$$[E_x Y] = \frac{A_e}{\delta} \int_0^\delta e^{-\Gamma z} \, e^{-jk_y y} \, d\tau. \tag{9.49}$$

y and z are the same linear functions in τ as in (9.33). We have omitted the explicit reference k to a specific test function to simplify the notation. After integration this becomes, by analogy with (9.34):

$$[E_x Y] = -\frac{A_e \, e^{-m_0}(e^{-\delta t} - 1)}{\delta t}. \tag{9.50}$$

Now we consider (9.17), the testing of the E_t field boundary condition. The expression for E_t in (9.17) contains normal and tangential derivatives of the longitudinal fields. The relation of these derivatives to the coordinate derivatives are given by (9.40). With this and (9.47) the E_t field can be rewritten as

$$E_t^-(y, z, k_y) = e^{-\Gamma z} e^{-jk_y y} \left[\frac{j\beta}{\gamma^2} t A_e - \frac{j\omega\mu}{\gamma^2} w A_h \right] \quad (9.51)$$

where t and w are given by (9.35) and (9.42). Testing with the triangular function (9.11) brings us to

$$[E_t Y] = \left[\frac{j\beta}{\gamma^2} t_1 A_e - \frac{j\omega\mu}{\gamma^2} w_1 A_h \right] \int_{-\delta_1}^{0} \left(1 + \frac{\tau}{\delta_1}\right) e^{-\Gamma z} e^{-jk_y y} d\tau$$

$$+ \left[\frac{j\beta}{\gamma^2} t_2 A_e - \frac{j\omega\mu}{\gamma^2} w_2 A_h \right] \int_{0}^{\delta_2} \left(1 - \frac{\tau}{\delta_2}\right) e^{-\Gamma z} e^{-jk_y y} d\tau. \quad (9.52)$$

Integrating by parts results in

$$[E_t Y] = \left[\frac{j\beta}{\gamma^2} t_1 A_e - \frac{j\omega\mu}{\gamma^2} w_1 A_h \right] \left[\frac{e^{\delta_1 t_1} - 1}{\delta_1 t_1^2} - \frac{1}{t_1} \right] e^{-m_0}$$

$$+ \left[\frac{j\beta}{\gamma^2} t_2 A_e - \frac{j\omega\mu}{\gamma^2} w_2 A_h \right] \left[\frac{e^{-\delta_2 t_2} - 1}{\delta_2 t_2^2} + \frac{1}{t_2} \right] e^{-m_0} \quad (9.53)$$

or

$$[E_t Y] = \frac{j\beta A_e \, e^{-m_0}}{\gamma^2} \left[\frac{e^{\delta_1 t_1} - 1}{\delta_1 t_1} + \frac{e^{-\delta_2 t_2} - 1}{\delta_2 t_2} \right]$$

$$- \frac{j\omega\mu A_h \, e^{-m_0}}{\gamma^2} \left[\frac{w_1(e^{\delta_1 t_1} - 1)}{\delta_1 t_1^2} + \frac{w_2(e^{-\delta_2 t_2} - 1)}{\delta_2 t_2^2} - \frac{w_1}{t_1} + \frac{w_2}{t_2} \right]. \quad (9.54)$$

Observe that, as in (9.44), for segments without a corner the last two terms w_1/t_1 and w_2/t_2 are equal and cancel each other out.

The testing of the H_x and H_t boundary conditions is performed in an analogous way as the testing of the E_x and E_t boundary conditions.

9.6 Inverse Fourier transformation

The spectral integration is performed in the classical way as described in Chapter 6. For $k_y \in [-k_c, k_c]$ we use simple Gauss quadrature integration formulas along a suitable complex integration path. As in the previous chapter the asymptotic part of the integration, that is, $k_y \in [k_c, +\infty[$ and $k_y \in]-\infty, -k_c]$, allows a special treatment. In this asymptotic part Γ is replaced by $|k_y|$ and the simplified asymptotic cascade is used to find the scattered fields. We will concentrate on the positive part, $k_y \in [k_c, +\infty[$, of

9.6 INVERSE FOURIER TRANSFORMATION

the asymptotic integration in which $\Gamma = k_y$. The other, negative, part is left to the reader.

From equations (9.34), (9.36), (9.39), and (9.44) it can be seen that the longitudinal incoming fields contain only terms proportional to

$$\frac{e^{-ak_y}}{k_y^2} \qquad \text{Re}(a) \geq 0. \tag{9.55}$$

For equations (9.34), (9.36), and (9.39) this is evident because t, u, w, Γ, m_0, and n_0 are all proportional to k_y when Γ is replaced by k_y. It is also the case for equation (9.44) because the last two terms cancel out. Indeed when we replace Γ by k_y we get

$$\lim_{\Gamma \to k_y} \frac{w_1}{t_1} = \frac{jk_y \sin \alpha_1 - k_y \cos \alpha_1}{jk_y \cos \alpha_1 + k_y \sin \alpha_1} = \frac{-k_y e^{-j\alpha_1}}{jk_y e^{-j\alpha_1}} = j \tag{9.56}$$

which is independent of α_1 and consequently equal to w_2/t_2. The common spectral asymptotic character (9.55) of all spectral longitudinal incoming fields is due to the special choice of the basis functions. If we had used also a piecewise linear representation for the transversal tangential fields we had terms proportional not only to $1/k_y^2$ but also to $1/k_y^3$. From equations (9.50) and (9.54) it can be seen, with the same reasoning as above, that the integration over a test function introduces an extra factor $1/k_y$. The fact that the observation integration always results in an extra factor $1/k_y$ is due to the special choice of the test functions.

First we concentrate on the asymptotic part of the incoming coefficients $[XY]_{kl}^{\text{in, as}}$ of the system matrix. From the above it is clear that these coefficients are the sum of terms of the following form:

$$[XY]_{kl}^{\text{in, as}} = C \int_{k_c}^{+\infty} \frac{e^{-ak_y}}{k_y^3} \, dk_y \qquad \text{Re}(a) \geq 0 \tag{9.57}$$

with C a function of β. We can rewrite (9.57) as an exponential integral [13]:

$$[XY]_{kl}^{\text{in, as}} = \frac{C}{k_c^2} \int_1^{+\infty} \frac{e^{-ak_c x}}{x^3} \, dx = \frac{C}{k_c^2} E_3(ak_c) \tag{9.58}$$

of which fast and accurate algorithms are available.

The scattered coefficients $[XY]_{kl}^{\text{sc, as}}$ also contain the $1/k_y^3$ spectral character originating from the excitation and observation integration. However, due to the solution of the layered medium an extra function $f(k_y)$ is introduced which describes the propagation and reflection in the layers. Hence $[XY]_{kl}^{\text{sc, as}}$

can be written as a sum of terms of the following form:

$$[XY]_{kl}^{sc,as} = \int_{k_c}^{+\infty} f(k_y) \frac{e^{-ak_y}}{k_y^3} dk_y \quad \text{Re}(a) \geq 0$$

(9.59)

$$= \frac{e^{ak_c}}{ak_c} \int_0^{+\infty} \frac{f\left(\frac{x}{a} + k_c\right)}{\left(\frac{x}{a} + k_c\right)^3} e^{-x} dx.$$

This last integral can be evaluated with Gauss–Laguerre quadrature.

As in the previous chapter $f(k_y)$ is a polynomial in β (and in γ^2 and $1/\gamma^2$) which means that we can evaluate (9.59) for each coefficient in the polynomial independent of β. Hence, the asymptotic integrations have to be done only once for each frequency in the iterative search for the propagation constant β.

9.7 Impedance calculation

9.7.1 Introduction

Until now we have concentrated on the calculation of the propagation constant β of the modes and on the corresponding modal E_x, E_t, H_x, and H_t field distributions on the conductor boundaries. The complete circuit model requires the impedance matrix associated with the modes. In Chapter 4 this impedance matrix is calculated from the mode current in each conductor, the powers propagated by each mode, and the cross-powers between the modes.

It was not necessary to make any assumptions about the material parameters of the conductors for the calculation of the propagation constant. This derivation was valid even for pure dielectrics. However, in the impedance calculation it is necessary to impose a restriction on these material parameters. This is necessary in order to give a meaningful interpretation to the total longitudinal current. This requires that the conductors are good conductors in which the displacement current is negligible as compared to the conduction current. The sufficiently high conductivity also makes it possible to restrict the power calculation to the layers and neglect the power propagated inside the conductors.

9.7.2 Longitudinal current

The total longitudinal current I_j on conductor j ($j = 1, \ldots, C$) is calculated from the contour integral of the transversal tangential magnetic field H_t

9.7 IMPEDANCE CALCULATION

around the conductor surface because the displacement current inside the conductor is neglected. In our approach the transversal tangential field is represented by H_t. If there are N_j segments on conductor j and if H_{t,j_k} ($k = 1, \ldots, N_j$) is the amplitude of the basis function of H_t on the kth segment then the total current on conductor j is given by

$$I_j = \oint_{c_j} \mathbf{H} \cdot \mathbf{dl} = \sum_{k=1}^{N_j} H_{t,j_k}. \tag{9.60}$$

9.7.3 Propagated power

The power propagated in sourceless layers, that is, in layers without conductors, is calculated in the spectral domain by the general technique described in Chapter 6. In the previous chapter the power in the source layers around the wires is calculated in the space domain by a two-dimensional numerical integration. In this chapter however the power in the source layers is calculated in the spectral domain.

Suppose we want to calculate the cross-power P_{fg} between mode f and mode g propagated in a source layer. From Chapter 6 it follows that this power is given by

$$P_{fg} = \pi \int_{-\infty}^{+\infty} \int_z [E_{y,f}(k_y, z) H_{z,g}^*(k_y, z) - E_{z,f}(k_y, z) H_{y,g}^*(k_y, z)] \, dz \, dk_y. \tag{9.61}$$

The z integration is over the source layer. To simplify the notation we omit the subscript referring to the source layer. All material-dependent functions have reference to the material parameters of this source layer. It will follow from the expressions of the field components appearing in (9.61) that the z integration can be performed analytically. However this analytical integration results in very tedious expressions which would require enormous amounts of CPU time to evaluate. Therefore we opted for a Gaussian quadrature numerical integration to perform the z integration. This means that we need to evaluate the integrand of (9.61) for some discrete z values.

Each field component in (9.61) consists of an incoming and a scattered part. The incoming part is due to the conductor parts located in the source layer and the scattered part is due to the scattered fields of all conductors. The incoming field components are expressed in terms of the longitudinal spectral incoming fields $E_x^{in}(k_y, z)$ and $H_x^{in}(k_y, z)$. These fields in turn follow from (9.19)–(9.24). The scattered field components on the other hand are expressed as functions of TE and TM field components in the spectral TE and TM cascades of the layered medium. Hence we get the following expressions, omitting the f and g subscripts, for the field components arising in (9.61):

$$E_y(k_y, z) = E_y^{in}(k_y, z) + E_y^{sc}(k_y, z)$$

$$= -\frac{j\omega\mu}{\gamma^2}\frac{dH_x^{in}}{dz} - \frac{k_y\beta}{\gamma^2}E_x^{in} + \frac{1}{k^2}(k_y E'^{sc} + \beta E''^{sc}) \quad (9.62)$$

$$H_z(k_y, z) = -\frac{j\beta}{\gamma^2}\frac{dH_x^{in}}{dz} - \frac{k_y\omega\varepsilon}{\gamma^2}E_x^{in} + H_z^{sc} \quad (9.63)$$

$$E_z(k_y, z) = -\frac{j\beta}{\gamma^2}\frac{dE_x^{in}}{dz} + \frac{k_y\omega\mu}{\gamma^2}H_x^{in} + E_z^{sc} \quad (9.64)$$

and

$$H_y(k_y, z) = \frac{j\omega\varepsilon}{\gamma^2}\frac{dE_x^{in}}{dz} - \frac{k_y\beta}{\gamma^2}H_x^{in} + \frac{1}{k^2}(k_y H'^{sc} + \beta H''^{sc}). \quad (9.65)$$

The TE and TM field components are easily calculated from the spectral longitudinal incoming field components and the solution of the layered medium.

For E_x^{in} and H_x^{in} in (9.62)–(9.65) we can use the expressions derived in Section 9.5.2. However these expressions are not valid when we want to know $E_x^{in}(k_y, z)$ and $H_x^{in}(k_y, z)$ at a z-coordinate located at the height of the excitation segment (region II in Fig. 9.7(a)). In this case the basis function integration has to be divided into a part above z in which $(z - z') > 0$ and a part under z in which $(z - z') < 0$.

First we calculate the contribution of an H_t field pulse basis function to $E_x^{in}(k_y, z)$. We consider the situation in Fig. 9.9 where $z_1 > z_0$; the opposite case where $z_0 > z_1$ is left to the reader. We divide the integration interval into a part $[0, d]$ where $(z - z') > 0$ and a part $[d, \delta]$ where $(z - z') < 0$.

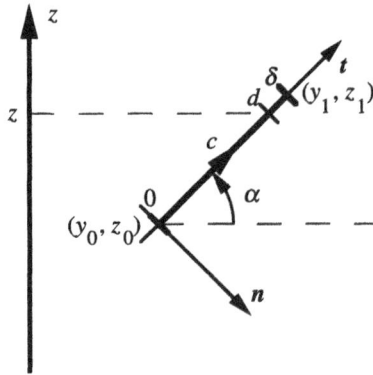

Fig. 9.9. Pulse function excitation segment with an observation point at the height of the segment.

9.7 IMPEDANCE CALCULATION

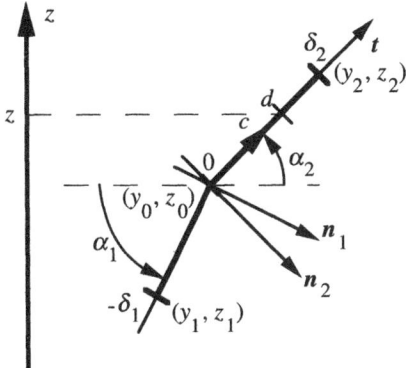

Fig. 9.10. Triangular function excitation segment with an observation point at the height of the segment.

The expression for $E_x^{in}(k_y, z)$, by analogy to (9.32), now becomes:

$$E_x^{in}(k_y, z) = -\frac{j\gamma^2 H_t}{4\pi\Gamma\varepsilon\omega\delta}\left[e^{-\Gamma z}\int_0^d e^{jk_y y'} e^{\Gamma z'}\,d\tau + e^{\Gamma z}\int_d^\delta e^{jk_y y'} e^{-\Gamma z'}\,d\tau\right]$$
(9.66)

with y' and z' given by (9.33). There is a sign difference between (9.32) and (9.66) because we are now explicitly concentrating on external incoming fields (cf. (9.5) versus (9.6)). After integration we get

$$E_x^{in}(k_y, z) = -\frac{j\gamma^2 H_t}{4\pi\Gamma\varepsilon\omega\delta}\left[\frac{e^{-\Gamma z + m_0}(e^{dt} - 1)}{t} + \frac{e^{\Gamma z + n_0}(e^{\delta u} - e^{du})}{u}\right]$$
(9.67)

with m_0, t, n_0, and u defined in (9.35) and (9.37). d is a function of z, given by

$$d(z) = \frac{z - z_0}{\sin\alpha}.$$
(9.68)

Now we can calculate the contribution of an E_x field triangular basis function to $E_x^{in}(k_y, z)$. We suppose that z is in the $[0, \delta_2]$ interval of the integration and that $z_2 > z_0$ as sketched in Fig. 9.10. The other configurations are left as an exercise. Thus the $[0, \delta_2]$ interval is divided in two parts $[0, d]$ and $[d, \delta_2]$. By analogy with (9.43) we get for $E_x^{in}(k_y, z)$:

$$E_x^{in}(k_y, z) = \frac{E_x}{4\pi\Gamma}\left[e^{-\Gamma z} w_1 \int_{-\delta_1}^0 \left(1 + \frac{\tau}{\delta_1}\right) e^{jk_y y'} e^{\Gamma z'}\,d\tau\right.$$
$$+ e^{-\Gamma z} w_2 \int_0^d \left(1 - \frac{\tau}{\delta_2}\right) e^{jk_y y'} e^{\Gamma z'}\,d\tau$$
$$\left.+ e^{\Gamma z} v_2 \int_d^{\delta_2} \left(1 - \frac{\tau}{\delta_2}\right) e^{jk_y y'} e^{-\Gamma z'}\,d\tau\right]$$
(9.69)

9 POLYGONAL TRANSMISSION LINE STRUCTURES

where
$$v_2 = -\Gamma \cos \alpha_2 + jk_y \sin \alpha_2. \tag{9.70}$$

After integration by parts this becomes

$$E_x^{in}(k_y, z) = \frac{E_x}{4\pi\Gamma} \left\{ e^{-\Gamma z + m_0} \left[\frac{w_1(e^{-\delta_1 t_1} - 1)}{\delta_1 t_1^2} + \frac{w_2(e^{\delta_2 t_2} - 1)}{\delta_2 t_2^2} \right. \right.$$
$$\left. + \frac{w_1}{t_1} - \frac{w_2}{t_2} + \frac{w_2}{t_2}\left(1 - \frac{d}{\delta_2}\right) e^{dt_2} \right] \tag{9.71}$$
$$\left. + e^{\Gamma z + n_0} \left[\frac{v_2(e^{\delta_2 u_2} - e^{du_2})}{\delta_2 u_2^2} - \frac{v_2}{u_2}\left(1 - \frac{d}{\delta_2}\right) e^{du_2} \right] \right\}$$

with $d(z)$ again given by (9.68).

Finally $E_x^{in}(k_y, z)$ generated by an H_x basis function needs to be calculated. For the configuration of Fig. 9.10 the reader may verify that this contribution is equal to

$$E_x^{in}(k_y, z) = -\frac{\beta H_x}{4\pi\Gamma\varepsilon\omega} \left\{ e^{-\Gamma z + m_0} \left[\frac{e^{-\delta_1 t_1} - 1}{\delta_1 t_1} + \frac{e^{\delta_2 t_2} - 1}{\delta_2 t_2} \right] \right.$$
$$\left. + e^{\Gamma z + n_0} \frac{e^{\delta_2 u_2} - e^{du_2}}{\delta_2 u_2} \right\}. \tag{9.72}$$

The contributions of E_x, H_x, and H_t to $H_x^{in}(k_y, z)$ are calculated in the same way as $E_x^{in}(k_y, z)$. Note that the final expressions (9.67), (9.71), and (9.72) contain only exponential functions in z multiplied by polynomials in z. This means that the integrand of (9.61) will also contain only exponential functions multiplied by polynomials in z. This proves that the z integration can be done analytically. But from the expressions (9.67), (9.71), and (9.72) one can guess to which complex expressions this would lead us.

9.8 Case studies

9.8.1 *Microstrip configuration*

As a first example we consider a thick microstrip (Fig. 9.11) with rectangular or trapezoidal cross-section on a substrate with thickness $h = 0.635$ mm and relative permittivity $\varepsilon_r = 9.8$. The thickness t is one tenth of the top width $w = 3.0$ mm of the strip.

Figure 9.12 presents the propagation characteristics of a perfectly conducting microstrip with rectangular ($\phi = 90°$) and trapezoidal ($\phi = 45°$) cross-section. The normalized propagation constants are found to be in

9.8 CASE STUDIES

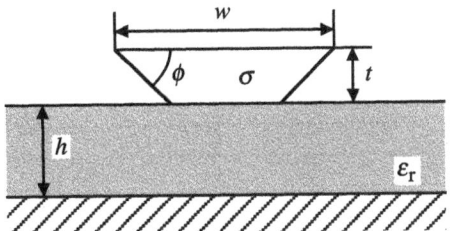

Fig. 9.11. A single thick microstrip.

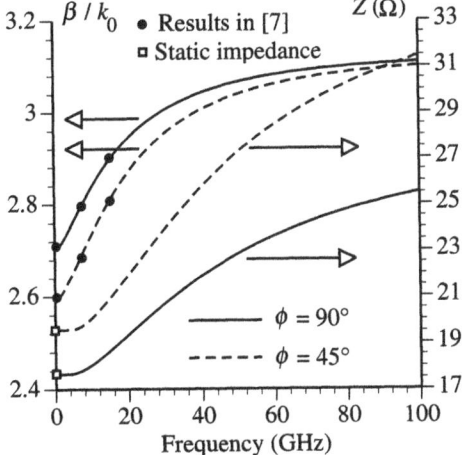

Fig. 9.12. Normalized propagation constant and impedance for the configuration of Fig. 9.11 when $\sigma = \infty$ for $\phi = 90°$ and $\phi = 45°$.

excellent agreement with the results of Michalski and Zheng in [7]. We verified the low-frequency limits of the impedances with results obtained by the capacitance program described in [4].

In Fig. 9.13, respectively Fig. 9.14, the complex propagation constants, respectively complex impedances, are shown when the strip of Fig. 9.11 has a conductivity $\sigma = 100$ kS/m. This corresponds to a bad conductor, worse than iron. As expected the attenuation constant ($= -\text{Im}(\beta)$) decreases with increasing frequency as a result of the skin effect. At low frequencies the real part of the propagation constant shows a small dip (see the enlargement on Fig. 9.13). This increase of $\text{Re}(\beta)$ with decreasing frequency is a result of the fact that the fields start to penetrate deep inside the microstrip with high complex permittivity $\varepsilon_r = \sigma/j\omega\varepsilon_0$. Indeed, at 1 GHz the skin depth is 50.3 µm and at 100 MHz it has already increased to 159 µm which corresponds to half the microstrip thickness. In this frequency range perturbation techniques [2] based on the skin effect are not applicable for this example. Note also

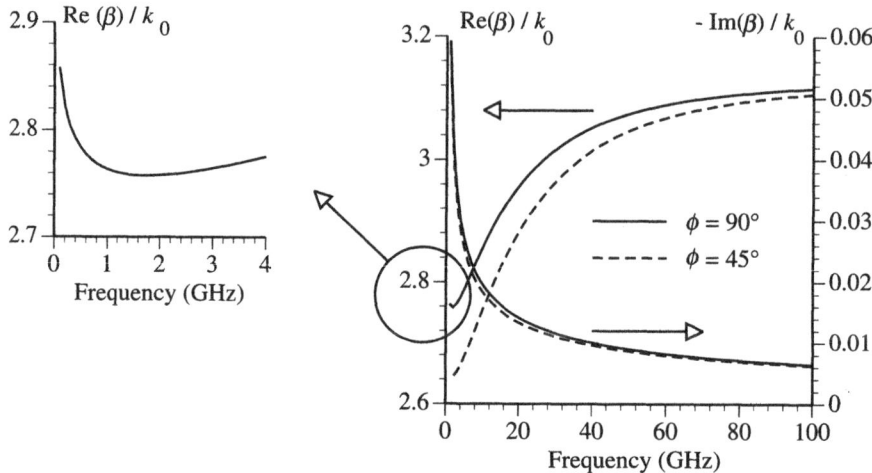

Fig. 9.13. Complex normalized propagation constant for the microstrip of Fig. 9.11 when $\sigma = 100$ kS/m for $\phi = 90°$ and $\phi = 45°$.

that $\mathrm{Re}(\beta)$ becomes almost equal to the propagation constant in the lossless case (Fig. 9.12) for frequencies over 8 GHz. The unbound character of $-\mathrm{Im}(\beta)/k_0$ at $f = 0$ owes its origin to the factor f from $k_0 = 2\pi f \sqrt{\varepsilon_0 \mu_0}$. The dispersion curves (Fig. 9.14) for the impedance show the same properties as in Fig. 9.13.

As a second example we took two perfectly conducting strips (Fig. 9.15) with $w = 1$ mm and $t = 0.3$ mm. Figure 9.16 presents the dispersion curves

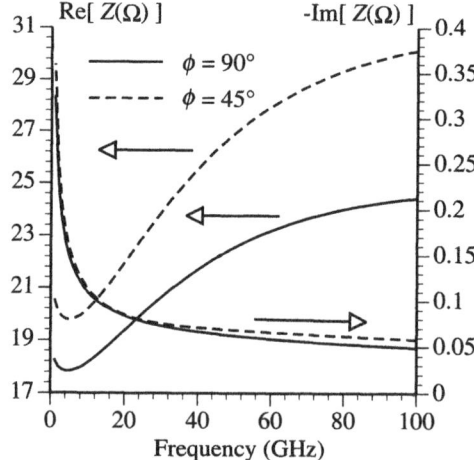

Fig. 9.14. Complex impedance for the microstrip of Fig. 9.11 when $\sigma = 100$ kS/m for $\phi = 90°$ and $\phi = 45°$.

9.8 CASE STUDIES   233

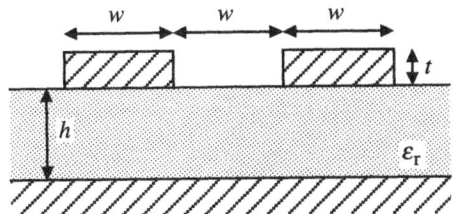

Fig. 9.15. Two coupled perfectly conducting thick strips.

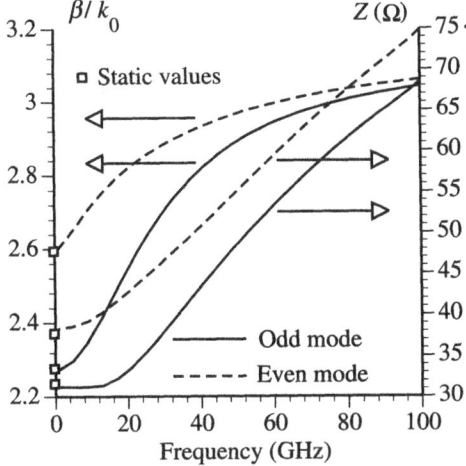

Fig. 9.16. Normalized propagation constants and impedances for the even and odd modes of the configuration of Fig. 9.15.

for the propagation constants and impedances of the even and odd modes. The statical values obtained with our capacitance program are also indicated.

9.8.2 Wire transmission lines

Figure 9.17 represents a wire with radius $a = 0.75$ mm approximated by a hexagon with the same area. The substrate has a thickness $h = 3$ mm and a relative permittivity $\varepsilon_r = 4$. H is the distance between the centre of the wire and the ground plane. In the simulations we used only two divisions on each side of the hexagon.

First we consider a perfectly conducting wire above the substrate with $H = 4.5$ mm. Figure 9.18 shows the propagation constant and impedance dispersion curves compared with results obtained in Section 8.10.1 for a round wire. Observe that the sudden increase of the impedance found in the

9 POLYGONAL TRANSMISSION LINE STRUCTURES

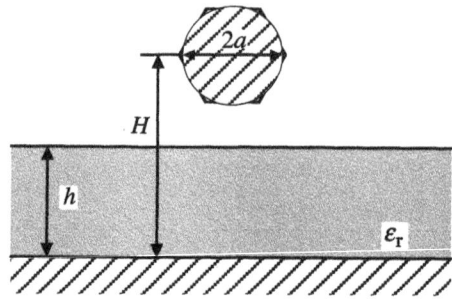

Fig. 9.17. A wire transmission line approximated by a hexagon.

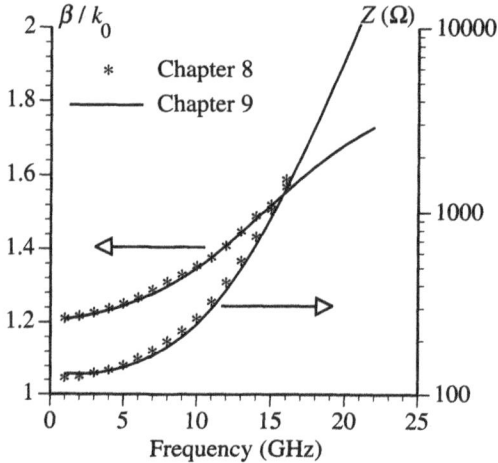

Fig. 9.18. Normalized propagation constant and impedance for the wire of Fig. 9.17 compared with results of the previous chapter ($\sigma = \infty$).

previous chapter is confirmed here. Slight differences in both impedance curves are probably due to the fact that in the previous chapter the power propagated in the source layer is determined in the space domain with a rather crude integration scheme. Figure 9.19(a), respectively Fig. 9.19(b), shows an arrow plot of the transversal electric, respectively magnetic, field components at a frequency of 2 GHz.

Now we take $H = 3$ mm such that the wire is semi-buried in the substrate (Fig. 9.20). This example was explicitly selected to illustrate our ability to handle conductors buried in two different layers. Figures 9.21 and 9.22 show the complex propagation constants and impedances as a function of frequency when the wire has a conductivity $\sigma = 100$ kS/m for two different values of the loss tangent of the substrate (tg $\delta = 0$ and tg $\delta = 0.001$). Note that the curves for the real part of both β and Z practically coincide for the

9.8 CASE STUDIES

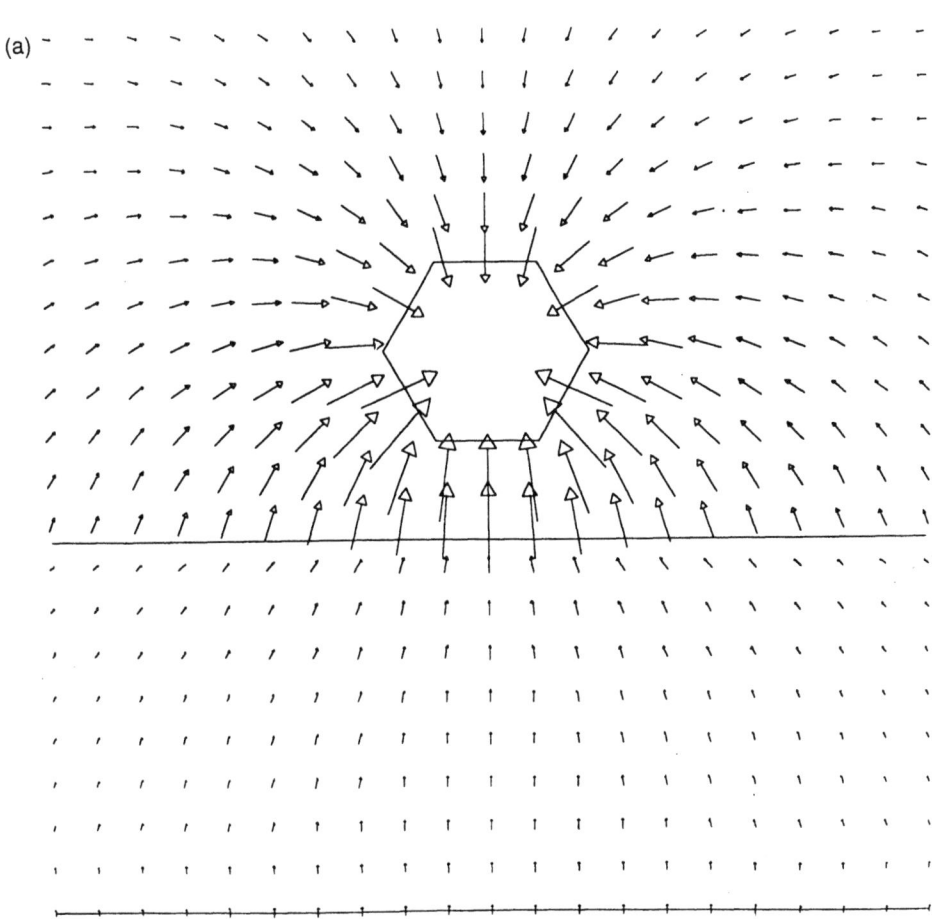

Fig. 9.19. Arrow plots of the transversal field components at 2 GHz. (a) The electric field.

Continued on page 236

236   9 POLYGONAL TRANSMISSION LINE STRUCTURES

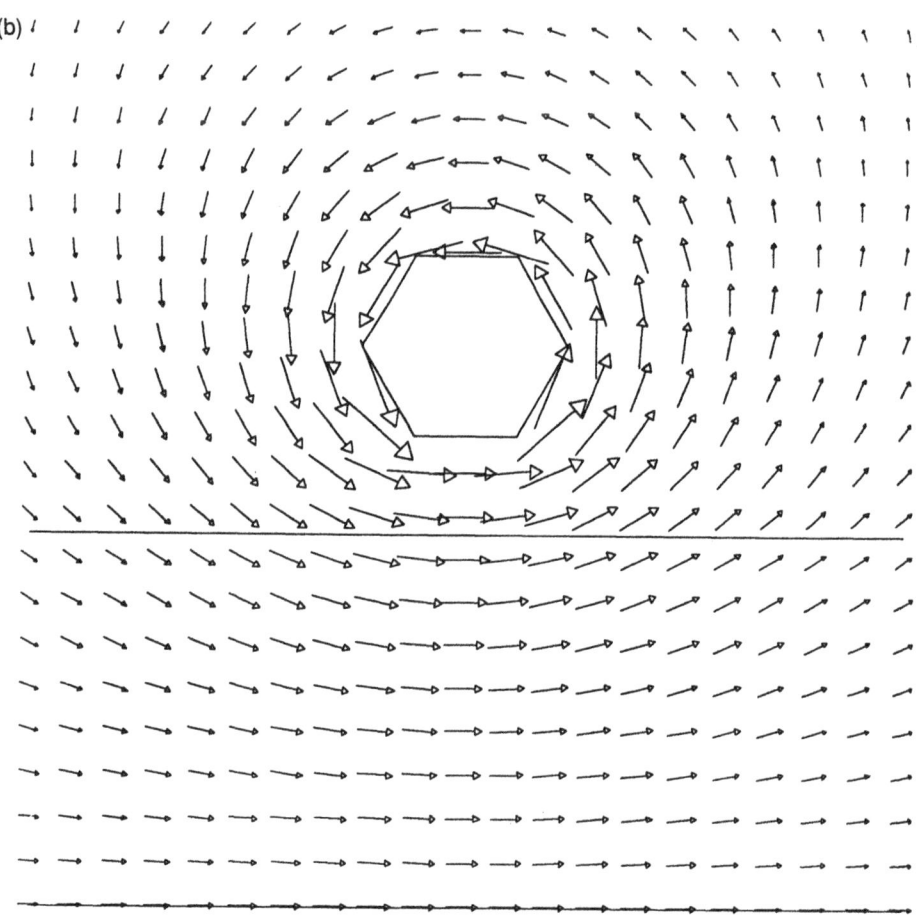

Fig. 9.19. (b) The magnetic field.

9.8 CASE STUDIES

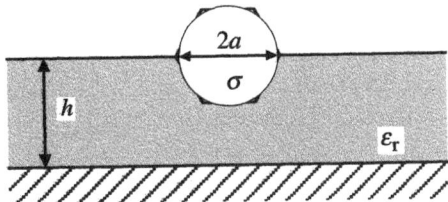

Fig. 9.20. A wire transmission line semi-buried in a substrate.

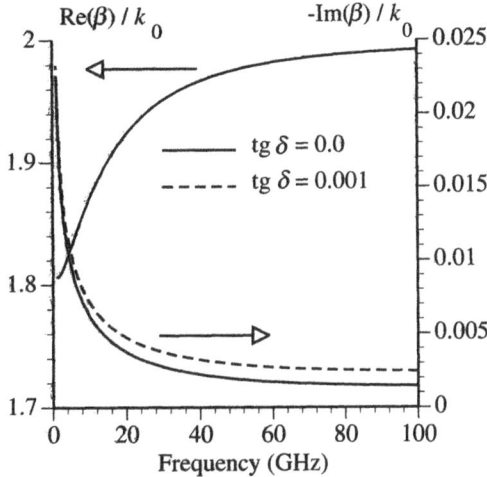

Fig. 9.21. Complex normalized propagation constant for the wire of Fig. 9.20 with $\sigma = 100$ kS/m for tg $\delta = 0$ and tg $\delta = 0.001$.

two values of tg δ. The imaginary part of Z (Fig. 9.22) shows the effect of losses in the conductor at low frequencies and the effect of losses in the substrate at high frequencies. Figure 9.23 shows the propagation constant as a function of the conductivity σ of the wire for a frequency $f = 10$ GHz and $f = 15$ GHz and a lossless substrate.

9.8.3 A multiconductor multilayer stripline

Figure 9.24 shows a three-layer stripline configuration in which three perfectly conducting microstrips are embedded ($h = 0.635$ mm, $w = 1.0$ mm, $t = 0.3$ mm, $H = 1.970$ mm, and $\varepsilon_r = 9.8$). The dispersion curves of the propagation constants of the three fundamental modes are shown in Fig. 9.25.

238   9 POLYGONAL TRANSMISSION LINE STRUCTURES

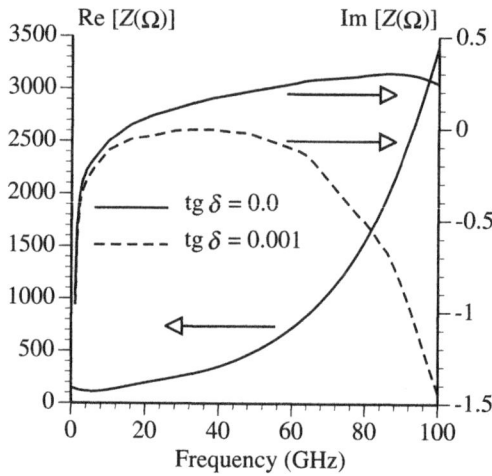

Fig. 9.22. Complex impedance for the wire of Fig. 9.20 with $\sigma = 100$ kS/m for tg $\delta = 0$ and tg $\delta = 0.001$.

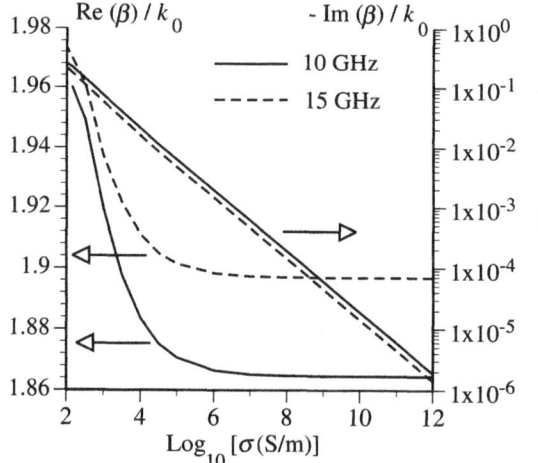

Fig. 9.23. Complex normalized propagation constant for the wire of Fig. 9.20 as a function of the conductivity σ when $f = 10$ GHz and $f = 15$ GHz.

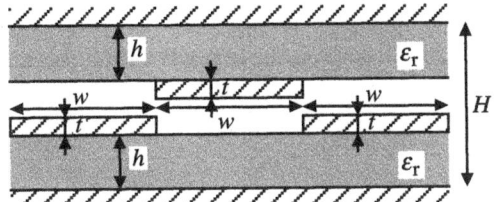

Fig. 9.24. A multilayer, multiconductor stripline.

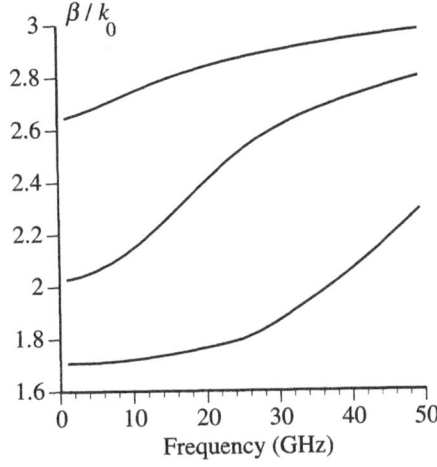

Fig. 9.25. Normalized propagation constants for the three fundamental modes of the stripline in Fig. 9.24.

References

[1] Wei, C., Harrington, R. F., Mautz, J. R., and Sarkar, T. K. (1984). Multiconductor transmission lines in multilayered dielectric media. *IEEE Transactions on Microwave Theory and Techniques*, **MTT-32**, 4, 439–450.

[2] Harrington, R. F. and Wei, C. (1984). Losses on multiconductor transmission lines in multilayered dielectric media. *IEEE Transactions on Microwave Theory and Techniques*, **MTT-32**, 7, 705–710.

[3] Delbare, W. and De Zutter, D. (1989). Space-domain Green's function approach to the capacitance calculation of multiconductor lines in multilayered dielectrics with improved surface charge modeling. *IEEE Transactions on Microwave Theory and Techniques*, **MTT-37**, 10, 1562–1568.

[4] Olyslager, F., Faché, N., and De Zutter, D. (1991). New fast and accurate line parameter calculation of general multiconductor transmission lines in multilayered media. *IEEE Transactions on Microwave Theory and Techniques*, **MTT-39**, 6, 901–909.

[5] Yang, J. J., Howard, G. E., and Chow, Y. L. (1991). Complex image method for analyzing multiconductor transmission lines in multilayered dielectric media. In *1991 International Symposium Digest IEEE Antennas and Propagation*, Volume II, pp. 862–865, London, Ontario.

[6] Michalski, K. A. and Zheng, D. (1989). Rigorous analysis of open microstrip lines of arbitrary cross section in bound and leaky regimes. In *1989 IEEE MTT-S International Symposium Digest*, Volume II, pp. 787–790, Long Beach, California.

[7] Michalski, K. A. and Zheng, D. (1989). Rigorous analysis of open microstrip lines of arbitrary cross section in bound and leaky regimes. *IEEE Transactions on Microwave Theory and Techniques*, **MTT-37**, 12, 2005–2010.

[8] Kolk, E. W., Baken, N. H. G., and Blok, H. (1990). Domain integral equation analysis of integrated optical channel and ridge waveguides in stratified media. *IEEE Transactions on Microwave Theory and Techniques*, **MTT-38**, 1, 78–85.

[9] Charles, J., Baudrand, H., and Bajon, D. (1990). A full-wave analysis of arbitrarily shaped dielectric waveguide using Green's scalar identity. *IEEE Transactions on Microwave Theory and Techniques*, **MTT-39**, 6, 1029–1034.

[10] Olyslager, F. and De Zutter, D. (1991). New boundary integral equation for waveguides with arbitrary cross section embedded in a multilayered dielectric. In *1991 International Symposium Digest IEEE Antennas and Propagation*, Volume II, pp. 866–869, London, Ontario.

[11] Olyslager, F., Blomme, K., and De Zutter, D. (1992). Full-wave eigenmode determination of propagation constants and impedances of coupled polygonal conductors in multilayered media. In *Proceedings of the 1992 URSI International Symposium on Electromagnetic Theory*, pp. 400–402, Sydney, Australia.

[12] Olyslager, F., De Zutter, D., and Blomme, K. (1993). Rigorous analysis of the propagation characteristics of general lossless and lossy multiconductor transmission lines in multilayered media. *IEEE Transactions on Microwave Theory and Techniques* (in press).

[13] Abramowitz, M. and Stegun, I. A. (1970). *Handbook of Mathematical Functions with Formulas, Graphs and Mathematical Tables*. Dover, New York.

APPENDIX A
INTEGRAL REPRESENTATION OF THE LONGITUDINAL MODAL FIELDS

In this appendix the integral representations used in Chapter 8, equation (8.5), and Chapter 9, equations (9.1), (9.2), and (9.9), will be derived together with some relevant relations between the longitudinal field components and the other field components that are also used in these chapters.

The results presented below, and more particularly the integral representations, form the starting point of the theory developed in Chapters 8 and 9. The derivation of these results is by no means new and can certainly be found in several textbooks. We simply want the reader to have the basic equations readily to hand.

Let's start from the sourceless Maxwell rotor equations in the frequency domain:

$$\nabla \times \mathbf{e} = -j\omega\mu\mathbf{h}$$
$$\nabla \times \mathbf{h} = j\omega\varepsilon\mathbf{e}. \tag{A.1}$$

Both the permittivity ε and the permeability μ can be complex. We now concentrate on the modal solutions of (A.1) for the general multiconductor lines considered in this book. Hence, we are looking for solutions which can be written as

$$\mathbf{e}(x, y, z) = \mathbf{E}(y, z)\,e^{-j\beta x} = (E_x \mathbf{1}_x + \mathbf{E}_t)\,e^{-j\beta x}$$
$$\mathbf{h}(x, y, z) = \mathbf{H}(y, z)\,e^{-j\beta x} = (H_x \mathbf{1}_x + \mathbf{H}_t)\,e^{-j\beta x} \tag{A.2}$$

where x is the propagation direction and β represents the (complex) propagation constant of the mode. We have separated out the longitudinal modal fields E_x and H_x from the transversal field components $\mathbf{E}_t$ and $\mathbf{H}_t$. In the following the x-components will be used as the basis functions from which all other field components can be derived.

Substituting (A.2) into (A.1) and separating out the longitudinal and transversal components yields

$$\gamma^2 \mathbf{E}_t = -j\beta\nabla_t E_x + j\omega\mu\mathbf{1}_x \times \nabla_t H_x$$
$$\gamma^2 \mathbf{H}_t = -j\beta\nabla_t H_x - j\omega\varepsilon\mathbf{1}_x \times \nabla_t E_x \tag{A.3}$$

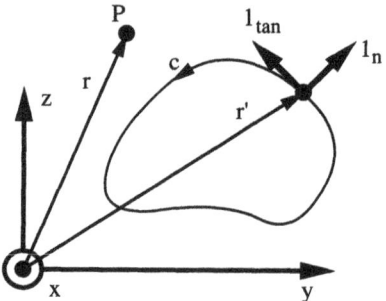

Fig. A.1. Two homogeneous regions separated by a boundary curve c.

and

$$\nabla_t \times \mathbf{E}_t = -j\omega\mu H_x \mathbf{1}_x$$
$$\nabla_t \times \mathbf{H}_t = j\omega\varepsilon E_x \mathbf{1}_x \tag{A.4}$$

where $\gamma^2 = \omega^2 \varepsilon \mu - \beta^2$ and $\mathbf{r} = y\mathbf{1}_y + z\mathbf{1}_z$. Combination of both sets of equations leads to the following Helmholtz type of equations for the longitudinal field components:

$$\nabla_t^2 E_x(\mathbf{r}) + \gamma^2 E_x(\mathbf{r}) = 0$$
$$\nabla_t^2 H_x(\mathbf{r}) + \gamma^2 H_x(\mathbf{r}) = 0. \tag{A.5}$$

To obtain (A.3), (A.4), and (A.5) we explicitly assumed that ε and μ are contants as a function of place. It is clear from (A.3) that the transversal components can be derived from the longitudinal ones.

We now consider two homogeneous regions separated by a boundary curve c (see Fig. A.1). The outward-pointing (with respect to the region inside c) unit normal vector on c is indicated by $\mathbf{1}_n$. The unit tangential vector to c is indicated by $\mathbf{1}_{\tan}$ and $\mathbf{1}_n \times \mathbf{1}_{\tan} = \mathbf{1}_x$. We will now show that E_x and H_x at any point P (located either inside c or outside c) can be derived from a knowledge of the fields and their derivatives on the boundary c.

We first concentrate on E_x and on a point P outside c as shown on Fig. A.1. The starting point of the calculation is the Helmholtz equation satisfied by E_x (see (A.5)) and the scalar Green's function $G(\mathbf{r}|\mathbf{r}')$ which satisfies

$$\nabla_t^2 G(\mathbf{r}|\mathbf{r}') + \gamma^2 G(\mathbf{r}|\mathbf{r}') = \delta(\mathbf{r} - \mathbf{r}')$$
$$G(\mathbf{r}|\mathbf{r}') = \frac{j}{4} H_0^{(2)}(\gamma|\mathbf{r} - \mathbf{r}'|) \quad \text{Re}(\gamma) \leq 0, \quad \text{Im}(\gamma) \leq 0. \tag{A.6}$$

$H_0^{(2)}$ represents the Hankel function of order zero and of the second kind. Applying Green's theorem to $G(\mathbf{r}|\mathbf{r}')$ and $E_x(\mathbf{r})$ in the region outside c,

INTEGRAL REPRESENTATIONS 243

immediately leads to the following expression:

$$E_x(\mathbf{r}) = -\oint_c \left[E_x(\mathbf{r}') \frac{\partial G(\mathbf{r}|\mathbf{r}')}{\partial n'} - G(\mathbf{r}|\mathbf{r}') \frac{\partial E_x(\mathbf{r}')}{\partial n'} \right] dc' \qquad (A.7)$$

indeed showing that $E_x(\mathbf{r})$ can be expressed as an integral involving E_x and its normal derivative at the boundary c. The integration extends over c and $\mathbf{r}'$ represents the position vector of a running integration point on c. Expression (A.7) remains valid when the observation point P approaches the boundary c provided the integral is interpreted in the principal value sense. In Chapters 8 and 9, however, Galerkin's method is applied. This means that the right-hand side of (A.7) is always multiplied by a testing function or weighting function. The integration over the testing function avoids the principal value problem as one would encounter it in the typical selfpatch calculations in the case of point matching or collocation.

In a completely analogous way an expression for $H_x(\mathbf{r})$ everywhere outside c can be derived:

$$H_x(\mathbf{r}) = -\oint_c \left[H_x(\mathbf{r}') \frac{\partial G(\mathbf{r}|\mathbf{r}')}{\partial n'} - G(\mathbf{r}|\mathbf{r}') \frac{\partial H_x(\mathbf{r}')}{\partial n'} \right] dc'. \qquad (A.8)$$

The values of ε and μ to be used in (A.1)–(A.8) are those belonging to the medium outside c.

For an observation point located inside c similar expressions to (A.7) and (A.8) can be found. It suffices to use the values of ε and μ belonging to the medium inside c throughout (A.1)–(A.8) and to delete the minus sign in front of the integrals in (A.7) and (A.8). The directions of $\mathbf{1}_n$ and $\mathbf{1}_{\tan}$ are left unchanged.

Once $E_x(\mathbf{r})$ and $H_x(\mathbf{r})$ are known, expressions (A.3) immediately lead to the other field components in terms of these longitudinal fields and their derivatives:

$$\begin{aligned} E_y &= -\frac{j\beta}{\gamma^2} \frac{\partial E_x}{\partial y} - \frac{j\omega\mu}{\gamma^2} \frac{\partial H_x}{\partial z} \\ E_z &= -\frac{j\beta}{\gamma^2} \frac{\partial E_x}{\partial z} + \frac{j\omega\mu}{\gamma^2} \frac{\partial H_x}{\partial y} \\ H_y &= -\frac{j\beta}{\gamma^2} \frac{\partial H_x}{\partial y} + \frac{j\omega\varepsilon}{\gamma^2} \frac{\partial E_x}{\partial z} \\ H_z &= -\frac{j\beta}{\gamma^2} \frac{\partial H_x}{\partial z} - \frac{j\omega\varepsilon}{\gamma^2} \frac{\partial E_x}{\partial y}. \end{aligned} \qquad (A.9)$$

The integrands of (A.7) and (A.8) depend on the longitudinal fields and their normal derivatives, that is, a set of four unknown functions. In Section

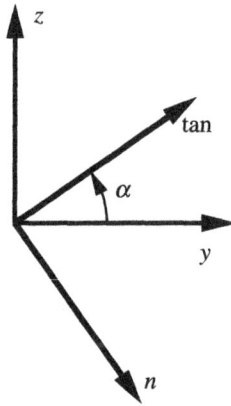

Fig. A.2. Cartesian coordinate system (y,z) and tangential coordinate system (n, tan) at a point on c.

9.3.4 it is proved that this set can be replaced by a new set of four functions E_x, H_x, E_{tan}, and H_{tan} together with their tangential derivatives along c. E_{tan} and H_{tan} are the projections of **E** and **H** on $\mathbf{1}_{\text{tan}}$. In this appendix we restrict ourselves to the derivation of the relations between the different field components that are needed for that proof.

Figure A.2 shows a cartesian coordinate system (y, z) and the corresponding (n, tan) coordinate system at an arbitrary point on c. The latter coordinate system is defined by the unit normal vector and the unit tangential vector at that point. For a regular point on the boundary c, the following relationships hold:

$$f_{\text{tan}} = \cos \alpha f_y + \sin \alpha f_z$$
$$f_n = \sin \alpha f_y - \cos \alpha f_z \tag{A.10}$$

and

$$\frac{\partial}{\partial n} = \sin \alpha \frac{\partial}{\partial y} - \cos \alpha \frac{\partial}{\partial z}$$
$$\frac{\partial}{\partial \tan} = \cos \alpha \frac{\partial}{\partial y} + \sin \alpha \frac{\partial}{\partial z} \tag{A.11}$$

where f_y and f_z are the cartesian components of a vector **f** and where f_{tan} and f_n are the corresponding normal and tangential components. Using (A.10) and (A.11) one can show that (A.9) leads to

$$E_{\tan} = -\frac{j\beta}{\gamma^2}\frac{\partial E_x}{\partial \tan} + \frac{j\omega\mu}{\gamma^2}\frac{\partial H_x}{\partial n}$$

$$E_n = -\frac{j\beta}{\gamma^2}\frac{\partial E_x}{\partial n} - \frac{j\omega\mu}{\gamma^2}\frac{\partial H_x}{\partial \tan}$$

$$H_{\tan} = -\frac{j\beta}{\gamma^2}\frac{\partial H_x}{\partial \tan} - \frac{j\omega\varepsilon}{\gamma^2}\frac{\partial E_x}{\partial n} \quad (A.12)$$

$$H_n = -\frac{j\beta}{\gamma^2}\frac{\partial H_x}{\partial n} + \frac{j\omega\varepsilon}{\gamma^2}\frac{\partial E_x}{\partial \tan}.$$

To conclude this appendix, (A.7) and (A.8) are reconsidered for the particular case of a perfectly conducting object. In this case E_x, $\partial E_x/\partial \tan$, and $E_{\tan}$ vanish on the conductor surface c. From the first equation in (A.12) it is clear that $\partial H_x/\partial n$ must also vanish. Consequently, for a perfect conductor, (A.7) and (A.8) take the form

$$E_x(\mathbf{r}) = \oint_c G(\mathbf{r}|\mathbf{r}')\frac{\partial E_x(\mathbf{r}')}{\partial n'}\,dc'$$

$$H_x(\mathbf{r}) = -\oint_c H_x(\mathbf{r}')\frac{\partial G(\mathbf{r}|\mathbf{r}')}{\partial n'}\,dc'. \quad (A.13)$$

These are the representations used in equation (8.5) and equation (9.9).

In this appendix we explicitly used the subscript 'tan' to make a clear distinction between the field components along that direction and the transversal field components, indicated by the subscript 't' in equations (A.1)–(A.6). In Chapters 8 and 9, the subscript 'tan' is replaced by 't' because there is no danger of confusion between the tangential and transversal directions in the context of these chapters.

AUTHOR INDEX

Abramowitz, M. 154, 158, 180, 182, 216–225
Ando, F. 137, 142, 159

Bajon, D. 204
Baken, N. H. G. 203
Baudrand, H. 204
Beyne, L. 100
Bianco, B. 49
Blok, H. 203
Blomme, K. 105, 204
Botte, M. 169
Brews, J. R. 6, 26, 29, 48, 83

Carin, F. 3, 6
Chang, F.-Y. 3, 12
Charles, J. 204
Chen, C. D. 98
Chow, Y. L. 203
Cohn, S. B. 10, 50
Collin, R. E. 59, 134, 198
Coombs, C. F. 98, 101
Criel, S. 12

Das, N. 137
Delbare, W. 101, 104, 137, 149, 169, 203
Denlinger, E. J. 137, 159
De Zutter, D. 3, 6, 12, 13, 53, 76, 84, 98, 99, 100, 101, 102, 104, 105, 137, 149, 151, 158, 169, 203, 204, 231
Dhaene, T. 3, 6, 12, 13, 76, 84
Djordjevic, A. R. 3

Faché, N. 6, 53, 76, 98, 99, 101, 102, 104, 137, 151, 158, 169, 203, 231
Felsen, L. B. 116, 117, 131, 148
Fukuoka, Y. 3, 137

Gardiol, F. 100
Geshiro, M. 6
Getsinger, W. J. 6, 49
Gu, Q. 6, 13, 83

Harrington, R. F. 3, 98, 101, 104, 141, 169, 203, 231
Horno, M. 137
Howard, G. E. 203

Itoh, T. 3, 96, 98–100, 114, 137, 142, 151, 159
Ibaragi, O. 101

Jansen, R. H. 1, 6, 9, 11, 50, 63, 71, 73, 114, 137, 161, 163

Kirschning, M. 6, 9, 50
Knorr, J. B. 137, 160
Kobayashi, M. 137, 142, 159
Kok, P. 169
Kolk, E. W. 203
Kong, J. A. 96

Lagasse, P. 76, 84
Lauer, R. B. 98
Lee, H. 6
Lindell, I. V. 6, 13, 83

McClintock, J. A. 3
Macdonal, B. M. 102
Magnus, W. 179
Marcuvitz, N. 116, 117, 131, 148
Marques, R. 137
Marx, K. D. 6
Mautz, J. R. 101, 104, 169, 203
Meixner, J. 99, 108, 141, 158
Messner, G. 101
Michalski, K. A. 104, 169, 203, 212, 231
Mittra, R. 114, 137, 142, 159

Nagai, N. 6
Neikirk, D. P. 3, 137

Oberhettinger, F. 179
Olyslager, F. 3, 76, 84, 101, 102, 104, 105, 137, 169, 203, 204, 231

Panini, L. 49
Parodi, M. 49
Peng, S. T. 98
Pitzalis, O. 1, 98
Pozar, D. 137

Ridella, S. 49

Sarkar, T. K. 3, 101, 104, 169, 203
Sawa, S. 6
Schelkunoff, S. A. 12, 30, 49
Schlafer, J. 98
Shibata, H. 101, 169
Silvester, P. 97
Soni, R. P. 179
Sphicopoulos, T. 100
Stanghan, C. J. 102
Stegun, I. A. 154, 158, 180, 182, 216
Sugita, E. 101

Terakado, R. 101, 169
Theodoris, V. 100
Tripathi, V. K. 6, 67, 137, 165

Tufekcioglu, A. 137, 160
Tzuang, C-K. C. 98

Van Hauwermeiren, L. 169
Van Hese, J. 98, 99, 137

Wang, J. 100
Webb, K. J. 3, 6
Wei, C. 101, 104, 169, 203, 231
Wiemer, L. 1, 6, 11, 73

Xu, Q. 3

Yagi, S. 6
Yang, J. J. 203

Zhang, Q. 3, 137
Zheng, D. 104, 169, 203, 212, 231

SUBJECT INDEX

The main entries in this subject index are listed alphabetically. Subentries are ordered logically, rather than alphabetically, to facilitate use.

basis functions
  thin strips 141
  wires 173
  polygonal-shaped conductors 211

case studies
  comparison between impedance
    definitions 50
  circuit model coupled hybrid modes 89
  thin strips 159
  wires 195
  polygonal-shaped conductors 230
characteristic impedance 9, 12
  single TEM mode 41
  single hybrid mode
    power–current definition 41, 46
    power–voltage definition 41, 43, 81
    voltage–current definition 41, 47
    other definitions 48
  matrix
    arbitrary coupled modes 85
  calculation
    polygonal-shaped conductors 226
circuit component models
  load 33
  driver 33
circuit current
  single TEM mode 41
  single hybrid mode 44, 46, 47
  two coupled hybrid modes 63
  arbitrary coupled modes 78
circuit description
  classical 4
circuit equations
  single TEM mode 37, 40
  single hybrid mode 37, 43, 46, 47
  two coupled hybrid modes 67
  arbitrary coupled modes 82
circuit line mode impedances 12
  two coupled hybrid modes 65
  arbitrary coupled modes 86
circuit model
  single TEM mode 40
  single hybrid mode 43, 46, 47, 48

  two coupled hybrid modes 63
  arbitrary coupled modes 78, 81
circuit principle 7
  single TEM mode 39
  single hybrid mode 42
  two coupled hybrid modes 63
circuit voltage
  single TEM mode 41
  single hybrid mode 44, 46, 47
  two coupled hybrid modes 67
  arbitrary coupled modes 79
conductor lines
  thin strips 97, 138
  wires 100, 170
  polygonal-shaped 103, 204

eigenmode, *see* mode
eigenvalue matrix
  thin strips 143
    matrix elements 155
  wires 178
    matrix elements 185, 189
  polygonal-shaped conductors 213
    matrix elements 214, 219
equivalent transmission line 5
  mode
    single TEM mode 27
    single hybrid mode 30
    two coupled hybrid modes 58
  impedance
    single TEM mode 28
    single hybrid mode 30
    two coupled hybrid modes 62

field distribution
  two coupled hybrid modes 68
  arbitrary coupled modes 84
field transmission line
  single TEM mode 20
  single hybrid mode 24
  two coupled hybrid modes 57
  arbitrary coupled modes 77

SUBJECT INDEX

Green's dyadic
  spectral
    thin strips 145
  spatial
    thin strips 149
  singular behaviour
    thin strips 155

impedance matrix 12
  arbitrary coupled modes 85
incoming fields
  wires 174
    spatial 179
    spectral 180
    angular decomposition 185
  polygonal-shaped conductors 217
integral equation 107
  thin strips 99, 139
  wires 102, 171
  polygonal-shaped conductors 105, 206
inverse Fourier transformation 116
  thin strips 149
  wires 190
  polygonal-shaped conductors 224

layered medium 112
longitudinal current
  polygonal-shaped conductors 227

Maxwell field equations
  space domain 115
    boundary conditions 115
  spectral domain 117
    boundary conditions 117
method of Galerkin
  wires 173
  polygonal-shaped conductors 211
method of moments
  thin strips 141
  wires 173
  polygonal-shaped conductors 211
modal field
  single TEM mode 17
  single hybrid mode 22
  two coupled hybrid modes 54
  arbitrary coupled modes 76
mode
  single TEM mode 17
  single hybrid mode 22
  two coupled hybrid modes 54
  arbitrary coupled modes 56

non-self patch elements
  thin strips 157

planar stratified medium, *see* layered medium
point matching
  thin strips 141
polygons, *see* conductor lines
power calculation
  sourceless layer 135
  wires 192
    layer with wires 193
  polygonal-shaped conductors 227
power distribution
  two coupled hybrid modes 68, 71
power equivalence principle 7
  single TEM mode 6
  single hybrid mode 28
  two coupled hybrid modes 59
  arbitrary coupled modes 80

reciprocity
  circuit model 11, 87

scattered fields
  wires 177
    spectral 187
    spatial 188
  polygonal-shaped conductors 218
self-patch elements 218
  thin strips 156
spatial fields
  sourceless layer 115
  inverse transformation of the spectral
    field 130
spectral field 113
  TE and TM modes 117
spectral transformation 116
surface waves 130

thin strips, *see* conductor lines
transmission line cascade
  spectral TE and TM modes 120
  general solution 124
    thin strips 146
    wires 177, 187
    polygonal-shaped conductors 218
  numerical considerations 125
  excitation
    thin strips 145
    wires 183
  transformation formulas
    spectral TE and TM modes 121
  asymptotic behaviour 128

wires, *see* conductor lines